Amateur Radio Mobile Handbook

Peter Dodd, G3LDO

Radio Society of Great Britain

Published by the Radio Society of Great Britain, Cranborne Road, Potters Bar, Herts EN6 3JE.

First published 2001.

© Radio Society of Great Britain, 2001. All rights reserved. No part of this publication may be reproduced, stored in a retrieval system, or transmitted, in any form or by any means, electronic, mechanical, photocopying, recording or otherwise, without the prior written permission of the Radio Society of Great Britain.

ISBN 1 872309 77 1

Cover design: Braden Threadgold Advertising.
Illustrations: Bob Ryan.
Typography: Ray Eckersley, Seven Stars Publishing.
Production: Mark Allgar.

Disclaimer
The opinions expressed in this book are those of the author, and not necessarily those of the RSGB. While the information presented is believed to be correct, the author, publisher and their agents cannot accept responsibility for consequences arising from any inaccuracies or omissions.

Printed in Great Britain by Black Bear Press, Cambridge.

Supporting web site: www.rsgb.org/mobilehandbook

Contents

Preface vii
Acknowledgements viii

Chapter 1: Going mobile 1

Mobile operation 1
A brief history of amateur mobile radio 2
Safety 5

Chapter 2: Mobile operating 9

HF operation 9
VHF/UHF operating 11
Repeater operation 12
APRS 15
Reference 19

Chapter 3: Installing radio equipment in vehicles 21

Introduction 21
Location of the transceiver 21
Connection to 12V supply 24
EMC 26
Further reading 29

Chapter 4: Mobile antennas 31

Inductive loading using a coil 31
Coil construction 33
The continuously loaded HF antenna 36
The W6AAQ continuous-coverage HF mobile antenna 39

The G3YXM Vaerial antenna.................................... 43
Matching the antenna to the feeder 44
VHF/UHF antennas 47
References 49

Chapter 5: Fixing an antenna to a vehicle.............. 51

Antenna location 51
Magnetic mount 51
Through-panel mounting 54
The chassis bracket 54
Luggage or roof rack antenna support 56
The hatchback antenna mount 58
The bull-bar antenna support.................................. 58
Coaxial cable antenna feeder 59

Chapter 6: Kite and balloon antenna supports 61

Kites 61
Balloons... 62
Choosing a site 63
Reference 64

Chapter 7: Bicycle HF mobile 65

The VE3JC bikemobile 67
The KB8U bikemobile 69
Note 70

Chapter 8: Maritime mobile operation................. 71

Installing HF radio equipment 71
Power supply connections..................................... 72
Antennas.. 72
Seagoing RF... 77
VHF... 78

Chapter 9: Mobile experimental activities 79

General antenna testing 79
DDRR mobile roof rack antenna 79
The toroidal antenna 82
An alternative feed for verticals 84
Other experimental work 85
References.. 85

Chapter 10: Walkabout mobile 87
The G0CBM walkabout mobile 88
The G3LDO walkabout mobile 90
FT-817 HF/VHF/UHF portable transceiver 91
Reference 97

Appendix 1: Resources 99
Antenna and equipment suppliers 99
Miscellaneous items 100

Appendix 2: Voice repeater lists 103

Index 113

Preface

There is nothing new about mobile operating, which has been around almost as long as amateur radio itself and has always been an enjoyable aspect of the hobby. Mobile and portable operation can offer an escape from EMC problems and indoor antennas, or for others the excitement of using a really good site or a DX callsign.

There have been considerable changes in both radio equipment and vehicle design since the last RSGB mobile book and this one has been written to address these changes.

The objectives of this book are:

- To provide practical advice on how to install radio equipment and antennas in a vehicle or boat so that interference to reception and electromagnetic interference to vehicle electronics are minimised.
- To overcome the problems of losses in HF antennas with suitable designs using coils or continuous loading.
- To give consideration to the location of the antenna, bearing in mind the construction of a modern vehicle and the proliferation of vehicle electronics.
- To advise on methods of operating amateur radio from a vehicle.

Other aspects of mobile radio are considered, such as maritime mobile, bicycle mobile and pedestrian mobile. In addition, the use of kite-supported or balloon-supported antennas and experimenting with unconventional antenna arrangements are described. VHF and UHF installations and operating, including the APRS system, are also presented.

General information such as radio equipment distributors, web pages and mobile references, together with repeater lists and maps are included in appendices.

I trust that this book will encourage radio amateurs to take up this fascinating aspect of the hobby. In spite of a greater use of electronics and the lack of space in modern vehicles, the availability of lightweight, comprehensive radios means that it has never been easier to become a mobile operator.

Peter Dodd, G3LDO
West Sussex, October 2001

Acknowledgements

I am indebted to the following people and organisations for assistance and material. Other sources of material are credited in the text.

Mike Dennison, G3XDV, recently RSGB Publications Manager, for permission to use RSGB material.

ARRL, for permission to use diagrams from early *Radio Amateur's Handbooks*.

Practical Wireless magazine, for permission to use photographs.

Waters & Stanton, for permission to use articles from their catalogue and for general help on mobile equipment and antennas.

Laurie Mayhead, G3AQC, for his help with obtaining information on maritime mobile operation and specifically on maritime antenna installations.

Bill Hall, G4FRN, for his loan of a book *A Guide to Small Boat Radio* and general help on maritime mobile matters.

Tony Walpole, VK6QG, for a description of his mobile station and photographs.

Dave Pick, G3YXM, for description and photos of his 'Vaerial' antenna.

Ian Keyser, G3ROO, for a description of his mobile station and photographs.

Peter Waters, G3OJV, for a description of his mobile station and photographs.

Charles Wilkey, G0CBM, for a description of his walkabout mobile station and photographs.

G0GCQ, for permission to photograph his mobile installation.

Chris Langmaid, for permission to photograph his maritime mobile installation.

CHAPTER 1

Going mobile

Mobile operation

A large number of radio amateurs around the world operate stations fitted into their cars. All bands from 1.8MHz to 1.3GHz are in use on AM, SSB, FM and CW. Mobile operating has particular attractions for those amateurs making regular car journeys to and from work, and most local mobile-to-mobile work is carried out on VHF/UHF FM. Other mobilers are HF enthusiasts, who have 100W SSB transceivers and loaded whip antennas fitted to their cars, and can make intercontinental contacts while driving.

The definition of what constitutes a mobile station is given below because there is often some confusion with some types of operating as to whether they should be mobile or portable. The *BR68 (REV8) September 2000* document issued by the Radiocommunications Agency defines mobile operation as follows:

"'Mobile' means located in the United Kingdom in any vehicle, as a pedestrian or on a vessel in inland waters. 'Inland waters' means any canal, river, lake, loch or navigation which is not a tidal water. The Licensee shall use the suffix '/M'.

'Maritime Mobile' means located on any vessel at sea (tidal water). The Licensee shall use the suffix '/MM'.

'Temporary Location' means a location other than the Main Station Address, in the United Kingdom, and in a fixed position. The Licensee shall use the suffix '/P' and every 30 minutes give the location to an accuracy of 5km by a generally used identifier.

The Licensee shall not establish or use the Station in an aircraft or other airborne vehicle.

No log need be kept in respect of Mobile and Maritime Mobile operations."

Like all rules they are open to some interpretation. Clearly, if you are operating from a moving vehicle or using a handheld, sign '/M'. If you are in the vehicle but stationary then you still sign '/M'.

If you are using an antenna similar to that shown in Fig 1.4 you would clearly not be able to drive very far. Similarly, if you use a kite or balloon to support the antenna this would limit mobility. What do you sign then?

My own interpretation is that if the radio is located in the vehicle, is powered from the vehicle supply, the antenna is fixed to the vehicle without any other support and the vehicle can be moved at all, then sign '/M'.

However, what if you are operating from a vehicle with a long wire antenna tied to a tree? According to the rules above then you would still sign '/M'. My interpretation is that if you are using the vehicle as a shack and you are using an external supply and/or the antenna is fixed to a support other than the vehicle, then sign '/P'.

As regards a manpack or handheld equipment: if the transceiver or pack is complete with its power supply and antenna then it is clearly mobile even if you set it down on a nearby wall or table to operate it. If you operate the manpack into a dipole fixed to a tree or some other fixed object then sign '/P'.

A brief history of amateur mobile radio

What follows are a few historical facts to place the mobile scene in perspective. My delving in the archives of the RSGB and Amberley museum libraries turned up enough interesting material to make up a book in its own right so I have to be selective and I will try to be concise.

Early days

Mobile amateur radio operation from a vehicle or a boat is an enjoyable aspect of the hobby we all take for granted today. However, when the first licences were issued after the second world war, mobile operation was not allowed. After much campaigning, mobile operation in the UK was finally permitted on 1 June 1954. A separate licence costing £1 was required in addition to the £2 for the main licence.

The first 'Mobile Column' appeared in the August 1954 edition of the *RSGB Bulletin* and it is surprising how much had been achieved in such a short space of time. The description of the G3MY mobile radio installation gives some idea of the technology of the time. His 80m transmitter comprised a 6C4 crystal oscillator with three crystals and a 2E30 PA running 12 to 15W. The AM modulator used a 9003 and an N78 with a crystal microphone. The HT was obtained from a 300V vibrator power pack. The receiver comprised a 6AG5 and a 6BE6 crystal-controlled converter with the car radio performing as a tuneable IF between 800 and 1100kHz. The antenna was a bumper-mounted 10ft centre-loaded whip, resonated for 80m with a coil wound with 16SWG wire on a 2½in former. The coil was fitted with a brass tuning slug which permitted the antenna to be tuned anywhere in the range from 3.6 to 3.8MHz; the design of this antenna, see Fig 1.1, came from *QST* and was reproduced later in the 1953 edition of the *ARRL Radio Amateur's Handbook*.

Also in this column is the description of one of the first stations to commence operation on VHF (144.64MHz) by G2ATK/M. The transmitter ran 10W using two 6C4s in push-pull and the receiver used double conversion.

CHAPTER 1: GOING MOBILE

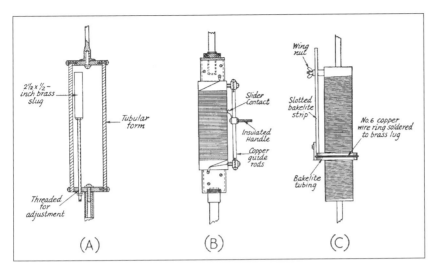

Fig 1.1. Three methods of varying loading-coil inductance, reproduced from the 'Mobile Equipment' chapter of the 1953 edition of the ARRL Radio Amateur's Handbook

The September 1954 issue of the *RSGB Bulletin* 'Two Metres and Down' column had a detailed description and photo of G5BM's VHF transceiver. This neat installation fitted into the glove compartment of a Ford Consul with a separate power supply and battery fitted into the boot of the car. In the same edition the 'Mobile Column', now edited by John A Rouse, G2AHL, was devoted mainly to equipment descriptions. This column and most subsequent ones contained discussions on how to obtain high voltages for valve equipment. Some used vibrator supplies but most used rotary converters; the majority of these were ex-military units.

Various items of ex-military radio equipment were pressed into mobile service. One popular item was the New Zealand ZC1 MkII transceiver. Although not exactly compact, measuring 22in by 10in by 10in, it was complete with a vibrator power supply. The frequency range of this transceiver was 2 to 4MHz and 4 to MHz in common with many other ex-army units. The lower scale could easily be tweaked to the top end of the 1.8 to 2MHz band so that it covered 160, 80 and 40m. An advertisement for this item is shown in Fig 1.2.

The February 1955 'Mobile Column' gave details of how to modify the ZC1 to improve the modulation level and cure the vibrator ripple problem. The September column was devoted to modifications on the ZC1, which included changes to the PA tank circuit to double the transmitter output power. This edition also gave details of the first 70cm mobile-to-mobile contact between G5KW/M and G8KW/M.

Mobile rallies

A mobile rally was first suggested by G5CV in the April 1955 'Mobile Column'. The first UK mobile rally took place at the Perch Inn, Binsley, Oxford, on 9 October 1955. More than 75 RSGB members and friends were there. The talk-in was on 160m, probably by the first car to arrive. Walter Blanchard, G3JKV, is the only radio amateur I know personally who attended that first rally and was able to give a flavour of what it

3

AMATEUR RADIO MOBILE HANDBOOK

The Walk-around Shop
has pleasure in offering you :—

TEST SET 87 incorporating mains Power Pack (as advertised in full on page 368 January issue). **£5.** Crg. 10s.

WIND FINDING ATTACHMENT for Air Speed Indicator. Comprising two small counters. Two Desyn-type follower motors (**Ideal for an antenna direction indicator**). Size of motors, 1¼in. long, 1in. diam., 6-way terminal block. Yaxley-type switch. Housed in metal outer case, fitted with plastic 360-degree dial. Price **8s. 6d.**, post paid.

BLOCK CONDENSERS.—8 μF, 600V W tropical. 750V W, normal price **5s.**, post 1s.

HEATER TRANSFORMERS.—6.3V, 1.5A output, 230V input. Price 7s., post paid.

TAPE SPOOLS. Clear Plastic, 1,200ft. Price **2s. 6d.**, post paid.

VALVES. 713A V.H.F. Triode. (Door Knob type) **9s.** p.p. GL446A Disc Sealed Triode (Lighthouse Tube) **25s.** p.p., 6SQ7 Double Diode Triode **7s.** p.p.

RHOMBIC AERIAL type 231, comprising: 2 reels of Copper Braided Aerial. Resistance Unit. Impedance Matching Unit (Variable tuning). Pyrex insulators, etc. **15s.** p.p.

RESISTANCE RECTIFIER UNIT. Bakelite case 2⅛in. x 2in. x 1½in. Containing ¼ wave selenium rectifier ¼ amp. 2 to 24 volts. **1s. 3d.** p.p.

(1) **AIR THERMOMETER** 1305. Grade 1 moving coil 3 mA meter movement.

(2) **BOOST GAUGE** (Barometric Capsule).

(3) **TURN AND SLIP INDICATOR.** Air operated gyroscope.
All three instruments for **9s. 6d.** p.p.

MAGNETIC SWITCH. 10 amp contacts, 12 to 24 volts. In bakelite case 4in. x 1½in. x 2in. RAF Ref. 5C/1722. **3s.** p.p.

WE HAVE AVAILABLE for callers only; a large selection of various types of meters.

NOTE: Orders and Enquiries to Dept. 'B'

PROOPS BROS. LTD., 52 TOTTENHAM COURT ROAD, LONDON, W.1

NEW ZEALAND TYPE ZC1 MARK II TRANSMITTER/RECEIVERS

TECHNICAL SPECIFICATION

The frequency covered is 2-4 and 4-8 Mc/s (37-150 metres). Power is obtained by means of a self-contained vibrator pack which operates with a 12 V. battery. Battery consumption is under 3 amps for the receiver, under 5 amps for the transmitter. Receiver valve line-up is 6U7G tuned R/F amplifier, 6K8G frequency changer, 6U7G I.F. amplifier, 6Q7G detector and audio amplifier, 6U7G output, 6U7G BFO. Transmitter valve line-up is P.A., Driver, M/O, amp, Pre/amp (osc.) utilising two 6V6G and three 6U7G. The I.F. frequency is 465 kc/s. Operation is on C.W., M.C.W. or R.T. Break-in operation is provided. AVC is incorporated on R/T. Two pre-set flick frequencies are provided and may be set to any frequencies within the tuning range.

Controllable pitch BFO, an efficient (switched) crash limiter, a moving coil meter checking both maximum output and battery voltage, are only a few of the refinements incorporated. The transmitter output is up to two watts, reliable communication up to about twenty miles may be obtained with a 12 foot whip aerial. This may be approximately doubled by using a 34 foot rod aerial. By utilising horizontal aerials and sky wave working, considerably greater range may be obtained. The complete unit is fully tropicalised and is fitted with a removable metal cover so that a watertight seal may be obtained. It is built into a substantial reinforced steel cabinet which is mounted on resilient mountings from which it is readily removable.

WE OFFER THESE NEW UNUSED UNITS
(Complete with valves)
at the remarkably Low Price of £6-19-6
(Plus 10/- carriage)

Available for use with above unit: Moving Coil Headphones, type F and Moving Coil Microphones, No. 7. Price for each 7/6d.

Shop hours 9 a.m. to 6 p.m.—Thurs.: 9 a.m. to 1 p.m
OPEN ALL DAY SATURDAY. Telephone: **LANgham 0141**

Fig 1.2. Advertisement for the New Zealand ZC1 MkII transceiver. Proops Bros Ltd was one of a number of shops in the area that sold all manner of ex-government radio equipment and components and was the cause of many radio amateur pilgrimages

was like. In those days there were no traders – just amateurs exchanging ideas about mobile radio. At this first rally neat home-made rigs made by G2HCG, G3XC, G5KW and G6AG aroused much interest. G3WW used a ZC1 mounted in the boot of the car, remotely controlled from the driving seat.

Another transmitter which was closely examined was W3WAM's 'Viking Mobile', covering all bands from 80 to 10m. This was one of the most popular transmitters used by mobile operators in the USA and was supplied in kit form. It provided 30 to 60W input (transmitters were rated by the power supply to the PA in those AM days) depending on the power supply voltage available.

In the June 1960 'Mobile Column' G2AHL reported that there were so many mobile rallies that it was no longer possible to give details of all of them. Not only were they more numerous, they were considerably larger, with over 1500 visitors checking in to the North Midlands Mobile Rally at Trentham Gardens, near Stoke-on-Trent, on 24 April 1960. They also became venues for obtaining RF components that became more difficult to find as the London radio emporiums, see Fig 1.2, closed down. While radio events like this have become larger, we have a way to go before they can rival those in Friedrichshafen or Dayton.

Commercial radio equipment

Nearly all the commercial equipment available for mobile use was of American manufacture. By the time UK amateurs

CHAPTER 1: GOING MOBILE

were permitted to use '/M' in 1954, the mobile scene was well underway in the USA.

The 1953 edition of *The ARRL Radio Amateur's Handbook* contained a 22-page chapter devoted to the construction of mobile equipment and antennas, and it contained several pages in the advertising section showing dealers and manufacturers selling mobile radio equipment and antennas. It was interesting to see just how far the development of mobile antennas had progressed, one of which is shown in Fig 1.3.

Early mobile operations

Although I was licensed in 1957 I didn't have a car until I went to live and work in East Africa in 1960.

My first attempt at a mobile rig was a crystal-controlled 5W 80m transmitter built in a PCR receiver. This rig was fine for short-range operating but not particularly successful for the longer ranges encountered in East Africa. Later I used a modified ex-army 19 set. In the 'sixties members of the Radio Club of East Africa provided communications for the East African Safari car rally, see Fig 1.4, and in my case the 19 set proved to be very successful on this venture.

Safety

Operational safety

It is important that when operating you do not take your eyes off the road, except for a glance at the frequency display or your log (see Chapter 2, Fig 2.1), in the same way that you glance in a driving mirror. While you are driving and not looking at the road you are driving out of control.

I once saw an incident, caught on a police car video, where a car slammed into the back

Fig 1.3. A mobile antenna design that could do with re-inventing. The loading coil inductance is selected by a contact at the end of a rod, which can slide up or down the inside of the coil. The contact is made to contact the selected turn of the coil by rotating the handle at the end of the rod. The spring-loaded whip above the coil allows the antenna to flex if it comes in contact with any obstruction and also allows the antenna to be stowed when the vehicle is garaged

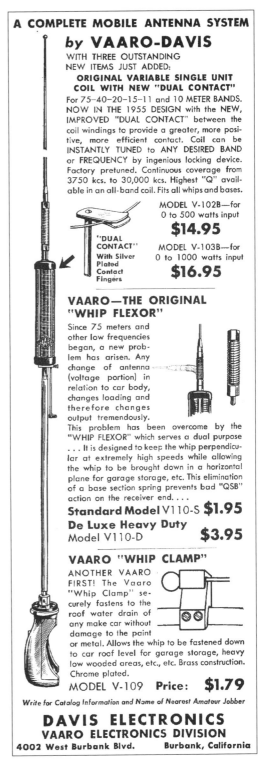

Fig 1.4. My mobile shack VQ4HX/M at one of the East African Safari control checks at Voi in Kenya, 1961. The antenna is a full-sized 7MHz vertical, which replaced the normal mobile antenna when operated from a fixed location. The car is an Armstrong-Siddeley Lanchester, with the roof raised to keep the operator cool

of a stationary vehicle. The driver was apparently trying to insert a tape into the cassette radio.

All equipment should be installed in such a way that in the event of accident or sudden braking it cannot injure the occupants of the car or cause a distraction to the driver. I have no problem with placing a transceiver on a passenger seat for occasional mobile operating (see Chapter 3) but it must be secure. If the rig moves during a driving manoeuvre it can cause a distraction. A friend of mine went to buy some eggs at a farm shop, which he placed on the passenger seat. On the way home he was aware that the eggs were slipping as he went round a hairpin bend. While trying to save the eggs he lost control of the car and plunged down an embankment. Fortunately he was not injured but the car was a write-off – and all the eggs were broken!

The difficulty of dictating operating safety requirements is that they have to be couched in terms that cover everything and everyone, regardless of abilities and situations. Some people are much better at doing more than one thing at a time than others. I know of at least two mobile operators who can send and receive 25WPM CW while driving. I find such a skill difficult to understand – it is probable that such operators can use CW just as easy as talking into a hand microphone.

I find that driving is a very automatic skill. If you commute the same journey every day it is as if the car seems to know its own way to work and back home again. On the other hand, if you are trying to find an address of a place where you have never been before, in heavy traffic, then the driving process requires full attention and concentration. Hence the stress of trying to operate with the talk-in station of a mobile rally you have never been to before, in a built-up area, in busy Sunday traffic.

You can be listening to the radio in a relaxed driving situation, when suddenly some traffic altercation ahead occurs. You will probably find that in this situation your mind switches off from the radio and on to driving the car in the changed circumstances; self-preservation kicks in.

For the last 30 years I have used a hand microphone when operating mobile, mainly because there was not much else, and I did become quite proficient at the juggling act necessary when changing gear. The use of a hand microphone is to discouraged in these days of high driving speeds and traffic density although driving an automatic vehicle does make life easier in this respect.

These days there are a lot of very nice lightweight boom microphones available, with or without the single earpiece (don't use a double-earpiece headset). I found the combination of boom microphone and VOX a pleasant operating experience – rather like talking to a passenger in the car. I have learned to suppress the unguarded 'remark' when someone cuts you up or pulls out in front of you.

Recent editions of the *UK Highway Code* state (rule 43) that "You must exercise proper control of your vehicle at all times. Do not use a hand-held telephone or microphone while you are driving. Find a safe place to stop. Do not speak into a hands-free microphone if it will take your mind off the road." The Department of Transport, in a letter to the Society, adds:

"The Highway Code is an advisory code of practice in that a failure to observe any of its provisions is not in itself an offence. Such failure, however, may be used in any court proceedings, which may arise. Current legislation already places the responsibility on drivers to have proper control of their vehicles at all times. A motorist who fails to do so as a result of distraction or lack of concentration is liable to prosecution".

Safety is a personal thing – know your limitations.

Batteries

It may not generally be realised that an explosion of considerable power can occur if a spark from a relay or other equipment ignites the gases given off by a lead-acid battery. C R Plant, G5CP, reported in the December 1962 'Mobile Column' of the *RSGB Bulletin* an alarming experience which took place when he was disconnecting a battery from a charging circuit. The power had been switched off and the vents firmly screwed into place when one of the large crocodile clips accidentally slipped and fell on to the top of the battery, thus short-circuiting two adjacent cells. A comparatively small spark occurred but this was immediately followed by a violent explosion; this blew out the end of the battery and splashed sulphuric acid around. Fortunately, in this case, apart from the loss of the battery, no serious damage resulted – it could have been a very different story if somebody had been standing close to the end that disintegrated! All lead-acid batteries give off an explosive gas mixture. Avoid any situation or power-circuit design (such as a relay), which may cause a spark in the vicinity of the battery, particularly when it is gassing due to charge or discharge.

A lead-acid battery can deliver 200 or 300A into a low-resistance load such as a short. This can cause a fire. Make sure all connections you make to a battery are fused at the battery end. Make sure that you work in such a manner that there is no danger of a tool being dropped so that it can fall on the terminals of a battery.

Carbon monoxide

In the same column G5CP noted the fate of a radio amateur who succumbed while working on his mobile radio equipment with the car engine running in an insufficiently ventilated garage.

Most people are aware that the fumes given out by a petrol engine are dangerous – the gas (carbon monoxide) is highly lethal, has no smell and so can creep up unawares if simple precautions are not observed. Never run an engine in a garage or other enclosed building unless you know that it has an adequate ventilation system and that it is working satisfactorily at the time. A better solution is to place the car rear outside the garage doorway so that the fumes are safely dispersed into the atmosphere.

CHAPTER 2

Mobile operating

HF operation

If you spend many hours behind the steering wheel of a car in the course of your business and you want to put some objectivity into your amateur radio operating then why not try DXing from your car?

Alan Birch, G4NXG, did just this and at the time of writing had worked over 327 countries; that's DXCC Honor Roll performance! However, the ARRL does not recognise DXCC from a mobile QTH so this achievement comes in the 'personal satisfaction' category.

In the early stages of operating HF mobile, nearly every country that you work will be a new one. Short-skip contacts to Europe will result in good reports and these can be often done while on the move.

Operating from a moving vehicle

The safety aspects of operating from a moving vehicle apply – see Chapter 1.

I find one of the most important considerations of operating HF while on the move is one of ergonomics: the ability to identify a control by touch and relative position without having to look at it so that I am not tempted to take my eyes off the road. It pays to have a transceiver that is reasonably ergonomic and simple to use.

A difficulty of operating HF mobile on the move is logging the QSO. Although keeping a log for mobile operating is not a legal requirement I do like to keep a record of my HF contacts. I tried using a small tape-recorder but found that there was too much equipment to operate at the same time.

Eventually I invented the steering wheel logbook shown in Fig 2.1. This comprises a postcard-sized piece of thick cardboard or plastic, with layers of paper attached, forming a small clipboard. This is fixed to the steering wheel with elastic bands. Holes and slots are cut in the plastic clipboard to enable it to be fixed quickly and conveniently. The contact information can be written on the top sheet of paper with hardly a glance at it. After the contact the layer of paper is then peeled off the top of the pack and thrown to the floor leaving a nice blank sheet for the next contact.

AMATEUR RADIO MOBILE HANDBOOK

Fig 2.1. The steering wheel clipboard logbook made from a postcard-sized piece of thick plastic material and small sheets of paper. This is fixed to the steering wheel with elastic bands using holes and slots cut in the plastic

At the end of the journey, all that has to be done is gather up all the bits of paper and enter the information into the logbook. Generally, for this type of operating, I am not concerned with new countries although some may emerge during an operating session.

A lot of time is spent listening. You may hear a short-skip station working a DX station. If the DX station is one you particularly want then it is often a good plan to look for a place where you can park the car so that you can concentrate on tail-ending the existing QSO in comfort.

Operating from a stationary vehicle

When it comes to searching for rare DX or contacting that rare DXpedition, the best way is to operate from a stationary vehicle. This allows you to select a good noise-free radio site and operate free of driving restrictions and vehicle electrical and traffic noise. Contacts can then be entered directly into the logbook.

Operating near the sea on a shoreline also appears to give improved DX potential. Not all promising-looking sites are good for HF operation. An example of one of these is a site near my home QTH, out in the countryside with very few trees – and an S8 noise background! It turns out that the noise is from the control and monitoring system for an underground reservoir.

When operating from a stationary vehicle you can fix a more efficient, but large, antenna that may be too large for mobile-in-motion operating – see Chapter 8. The only concern regarding stationary mobile operating is the state of the car battery – I know of at least one mobile operator who was not able to start his car after such a session! A voltmeter to monitor the battery voltage can be a useful accessory if you do a lot of static DX operating.

The secret of success in this type of operation is lots of listening and patience. Enter the frequencies of all interesting QSOs into the rig's local memories. You can then tune the band for any new activity (DX stations calling CQ), then scan through the local memory frequencies – much the same as operating from the home station.

An essential requirement for a mobile rig is split-frequency operation because many DXpeditions use this method of operating. Most rigs have this facility these days.

A VHF/packet system operating into a local DX cluster will be an advantage. This can be built using a handheld, a TNC and a laptop or handheld computer.

VHF/UHF operating

DX operating on VHF/UHF is done on the lower sections of the bands on CW and SSB and from a fixed location – normally an elevated QTH. The operating procedure is very similar to DX operating from a fixed site on HF, as described above. 50MHz is also becoming increasingly popular with mobilers because of possible long F2 or sporadic-E DX contacts when conditions are good.

VHF/UHF openings or lifts can be quite exciting, and a mobile QTH can be useful for taking advantage of them. Do have some method of working out your locator; this is much easier than trying to explain a remote QTH. A GPS unit is useful in this respect.

Most mobile on-the-move working takes place on the FM sections of the VHF/UHF bands.

In the early days of VHF mobile operating the procedure was dictated by equipment constraints. Crystals were expensive so stations tended to transmit only on one or two frequencies and tune around on a VFO receiver for a reply. This mode of operating was very inconvenient for mobile operating and VHF FM sections of the VHF/UHF bands became channelised, which was a considerable safety and convenience advantage for the mobile operator. Instead of fine-tuning a VFO dial, the operator was able to click a rotary control round to change frequency, and with practice this could be done without taking the eyes off the road.

The channels were spaced 25kHz apart and, to avoid the necessity of giving long strings of digits when specifying frequencies on the air, they were numbered. Channel 0 on the 144MHz band corresponded to 145.000MHz, while channel 1 was 145.025MHz and so on. This put the calling channel of 145.500MHz as S20. Many radios had the frequency dial calibrated in channel numbers.

Now the channel spacing has changed to 12.5kHz and so the channel numbering system has also had to be changed – there are twice as many channels so the calling frequency is channel 40.

This has led to some confusion and the channeling number system has tended to be used less these days, with the channel spelt out in terms of the last three digits of the frequency. Most VHF/UHF transceivers feature memories, which can be programmed and selected by a single control.

The normal way to set up a simplex contact while operating VHF/UHF mobile is to make (or answer) a CQ call on the mobile calling channel and to change frequency as soon as contact is made to one of the 'working' channels.

Using the standard 10 to 25W transceiver and a $^5/_8$-wave whip it will usually be found that the range for mobile-mobile simplex work is very unpredictable in low-lying urban areas and that mobile 'flutter' is a problem, particularly when both stations are on the move. Of course, one solution is to fit an add-on RF amplifier to boost the transmitter power, and perhaps also a preamplifier to improve the receiver sensitivity. Although these measures can give a useful increase in range, the extra expense, spectrum pollution and power consumption involved have led to the use of repeaters for mobile-to-mobile contacts.

Repeater operation [1]

Repeater stations are unattended installations on good radio sites such as hilltops or high buildings that relay amateur transmissions to provide wide area coverage for stations that might otherwise have restricted range. The installations are provided by amateurs generally as part of a group or club for the benefit of all licensed amateurs. In the UK most of the populated areas are within the 'service area' of at least one repeater. Repeater operation is most popular on the 2m and 70cm bands using NBFM.

In addition there are repeaters on 29MHz, 50MHz and 1.3GHz – see Appendix 2.

Repeater technology

The transmitter and receiver are conventional high-specification units. Generally a common receive and transmit antenna is used and the receiver and transmitter are isolated from each other by a frequency 'split' and very narrow band-pass filters as part of a combiner unit. Silver-plated cavity filters with high Q factors are most common, providing at least 60dB rejection in each path, although other solutions are just as successful. Sometimes separate transmit and receive antennas can be used but this introduces non-reciprocity of coverage between transmit and receive service areas.

The logic unit provides a number of functions. The repeater's transmitter is keyed by this unit which detects a valid access tone on the receiver's input frequency. This access tone can either be an audible tone (*toneburst*) lasting about half a second at a frequency of 1750Hz or a *continuous tone-coded squelch system* (CTCSS) tone which is a sub-audible tone continuously injected on the user's transmission. Not all repeaters use CTCSS but there are considerable technical advantages with this system, particularly when the repeater is co-sited on a busy radio site with potential interference sources. Once the repeater is 'accessed' it will relay the input frequency signal for a preset time or until the input carrier is dropped. After a short 'courtesy' gap to allow any other user to call in, an 'invitation to transmit' tone (often a Morse code 'T' or 'K', or a pip tone) is given, whereby another station can reply. The logic timer is reset until the next cycle. If the repeater receives no further input signal the transmitter is de-keyed after a short interval. A block diagram of a repeater is shown in Fig 2.2.

Because the repeater is available to all amateurs it is often necessary to restrict the time for each 'over' and the logic provides a *time-out* period which varies depending on the desire of the operating group from one minute up to half an hour.

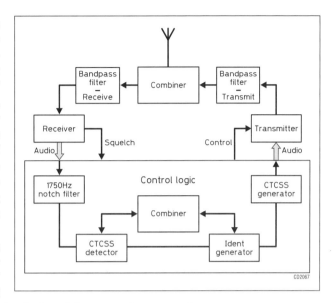

Fig 2.2. A simple block diagram of a repeater

The logic also prevents the transmitter from being continually keyed if a stray interfering signal appears on the input frequency. Reactivation of the cycle is only possible thereafter on receipt of a valid toneburst.

It is necessary for the repeater to periodically identify itself using Morse code at a speed of about 12 words per minute and the logic unit provides this function. When not in talk-through mode most repeaters send a short identification every 15 minutes or so. If CTCSS is in use this is also sent as a Morse character corresponding to the tone frequency after the repeater callsign. It is also permissible to use additional voice identification.

There is no restriction as to other functions provided by the logic unit. Some in use on UK repeaters include indication of over-modulation of the input signal, frequency high or low, or if the repeater is on standby power if the site mains supply fails. To avoid the repeater being kept open by an unmodulated carrier, modulation detectors are popular.

Repeater frequencies

Because of the need to share the available bandwidth with other amateur users, repeater operation is restricted to sub-bands with standard frequency splits. These sub-bands are then split into channels and it is necessary to specify operating parameters in order to avoid co-channel interference. Depending on the band in use, channel spacings of 10, 12.5 and 25kHz are in use in Europe on VHF and UHF. Channel designators are commonly used – see the tables of repeaters in Appendix 2.

Frequencies for repeaters are chosen to prevent where possible ground-wave interference between units. The nature of amateur radio and the restricted number of channels available means this is not always possible.

Techniques of service area pattern tailoring using cardioid antenna patterns and appropriate power limitations are sometimes necessary.

Often the 'ideal' site for a desired service area is just not available and compromises need to be made. The cost of renting high sites has increased markedly in the last few years. One solution being developed by some groups is to use clusters of linked repeaters to provide the desired area coverage. This is so especially on 144MHz where, due to site rental increases, many of the high-sited units had to close down. The introduction at the beginning of 2000 of 12.5kHz channel spacing on 144MHz in accordance with IARU recommendations facilitates new units providing equivalent coverage.

Using a repeater

There is normally only one repeater in an area on any particular band and it can become very busy. It is therefore necessary to make transmissions as short as possible to enable the best utilisation. The golden rule when operating via (or 'through') a repeater is the same as operation on other modes – listen before attempting to transmit! Listening to users on the output frequency will usually give a good indication of the protocol to adopt.

If the repeater is not in use, set your transmitter to the required input frequency together with a 1750Hz toneburst (or the correct CTCSS tone set) and make a short transmission in the form of "This is G3ZVW Mobile listening through GB3FN." Some repeaters need some five seconds of audio in which to enter repeat mode. When you release the transmission there will be a short period of unmodulated carrier followed by a tone or Morse character inviting others to reply.

Stations replying will use the format "G3YMK Mobile, this is G3TZM Portable in Harrow." Again the 'invitation to transmit character' should be waited for after the 'courtesy' gap, after which the contact proper can commence. Remember, there will be a time limit (timeout) so keep the reply brief and speak clearly. There is no need to repeat both callsigns but you ought to identify yourself by saying "From G3YMK" at the beginning and end of each transmission. Other stations may well join the QSO and will call in during the courtesy gap. If all stations can hear each other on the input frequency it is good manners to suggest moving to another frequency where a simplex QSO can take place, hence leaving the repeater for others to use. Remember that voice repeaters are listened to not only by radio amateurs but also many others, and it may be that listening to a voice repeater is the first introduction to amateur radio for many people. Local repeaters are very often the on-air meeting points for local amateurs and operation via them is a lot of fun.

Repeater abuse

Regrettably there are individuals who see repeaters as targets for abuse. It is a fact of life that the input to a repeater can be blocked easily and this denies the facility to legitimate users. Such behaviour, along with use of abusive language and obscenities, is heavily punished if the perpetrator is caught. The Radiocommunications Agency regularly monitors repeater output, and techniques of transmitter fingerprinting have led to numerous successful prosecutions. Whilst sometimes it is hard to ignore such abuse, repeater users are strongly advised to do so and not respond in any way.

APRS

If you are interested in the high-tech side of mobiling then this is for you. You will be able to make use of your VHF/UHF rig, computer (with some APRS software), TNC and GPS system all at the same time. The following is based on an article [1] written by Ciemon Dunville, G0TRT.

APRS stands for 'Automatic Packet Position Reporting System', and is the brainchild of Bob Bruninga, 34APR. The original idea was a track- the location of a fixed or mobile sta- of the location and general informa- ause it runs on a computer, the need duced. Of course, the use of APRS is ons – it can be used for events (talk- d day-to-day activities.

that it uses unconnected communi- e-to-one. For example, to pass a one- ation B to station C on a connected ith an unconnected system it takes duction of on-air time makes for a he Bulletin Board System (BBS), there used correctly, reliable long-range multiple hops. This gives a truly all-

one transmission so, to keep trans- m and thus ensure maximum effi- ation over the required area, is quite you realise that APRS can be used mobile position; four types of mes- power, height and gain etc. n one 256-byte packet but each one cognised, on-air protocol. Here are

```
= 5207.40N/00057.06W -PHG3260
```

This says that the station is located at latitude 52° 07.40" North, longitude 000° 57.06" West. It is showing a house icon on the map; it has the ability to send and receive messages; it is using 10W into an omni-directional antenna which has 6dB gain and is at 40m above average terrain.

Weather report

```
!041230/5207.40N/
00057.06W_c000s015g025t045
r100p200P150h40b10132
```

This says that the weather station is located at 52° 07.40" North, 000° 57.06" West. The system is an Ultimeter200 weather station and has no messaging capability. The weather report is timed at 1230 on the 4th of the month. The wind is from the north at 15 knots, gusting to 25 knots. The temperature is 45°F. Rainfall: 1 inch in the last hour, 2 inches in the last 24 hours, 1.5 inches of rain since midnight. The humidity is 40% and the pressure is 1013.2mb.

What can APRS do?

With APRS, the following can be sent:

- *Beacons* – your position, be it fixed or mobile. It is possible to indicate the type of station (anything from a Scout to a space shuttle including a car, bike, truck, school – the list is long).
- *Course and speed* – for dead reckoning.
- *Power, height and gain* – for frequency co-ordination by plotting circles on the map for each station.
- *DF report* – for foxhunting or interference location.
- *Weather* – including rain, pressure and humidity. Various weather stations can be linked directly to APRS to automatically send the information.
- *Objects* – it is possible to send object reports for such things as accident locations, rallies, meetings, and the list is limited only by your imagination.
- *Status* – used to inform of the station's current activity or something similar. A lot of stations put their best on-frequency DX here, others use it to put a point of contact, be it a BBS, phone number or e-mail address.
- *Messages* – All APRS messages are instant – as soon as you press 'send', it goes. The beauty of these messages is that they are in real time and as such it is possible to have large round-table discussions with as many stations as you like. It is also possible to send *bulletins*, which are messages that are transmitted a couple of times an hour for a day. These might be used to notify people of a large accident that has blocked a major road. *Announcements* can also be made – these are long-term messages, transmitted every couple of hours for a number of days; they could be used to let users know the local standards or point of contact for information.

CHAPTER 2: MOBILE OPERATING

Fig 2.3. APRS in abundance – Kenwood TH-D7E, Garmin Street Pilot GPS, Kenwood TM-D700E

- *Queries* – There are various queries that you can ask of another station, giving status, position, stations heard etc.

- *DXCluster* – APRS is very useful for the *DXCluster* user. By listening on the cluster frequency, APRS picks up spots and plots them on a world map: all the spot information is available on various pages and the beauty is that you never have to transmit, keeping cluster frequency loading to a minimum. Some software can be used to retune the radio and turn the beam automatically.

All this makes APRS is a very capable and flexible system.

What do I need?

There are several ways of setting the system up; all you need is a packet station, which includes a computer, TNC and a radio. Of course, operation can be achieved using Baycom boards or sound cards or using the latest radios that have TNCs and APRS built in, but they are not essential. An example of an APRS station is shown in Fig 2.3.

So, you've got your TNC and computer connected up and running at 1200 baud on 144.800MHz (the UK's dedicated APRS frequency). It is possible to use an ordinary terminal program to see the packets and decode them manually; it is an excellent way to learn about APRS. To appreciate the finer points of APRS, you need some APRS software. Without doubt, the most widely used software in the UK is UI-View written by Roger Barker, G4IDE, but there is software available also for the DOS, Windows®, WinCE®, Linux®, Macintosh®, PalmOS® operating systems and probably more. This is all available via the Internet and some links are given at in Appendix 1.

Trackers

Mobile stations have already been mentioned; these are commonly known as *trackers*, as they enable you to track them. A tracker is made up of a GPS receiver, a TNC and a radio. Once again, there are various types of set-up:

Fig 2.4. The Garmin Street Pilot – typical of the range of systems available with NMEA output

- *GPS* – This is a receiver which provides National Marine Electronics Association (NMEA) data for location information. Many receivers are available but, to be used with a tracker, it must have a data port and the ability to send NMEA data through that port. Pharos is a very small GPS receiver, just a little bigger than a box of matches, but it has no display, so this is not particularly useful if you want to be able to see information. At the other end of the scale, the Garmin Street Pilot, see Fig 2.4, is a top-of-the-range system that gives you a street map and as much information and data you could possibly want.
- *TNC* – Any TNC used as a tracker must have GPS parameters otherwise it cannot be used. The Picopacket is a very small and capable system that is widely used for small, lightweight set-ups.
- *Radio* – In a developed APRS network (the UK net is still very young) a handheld radio is all that is needed but, while the network here is developing, any mobile radio will do; obviously, greater power will enable you to be tracked at greater ranges. The advent of the Kenwood TH-D7E and TM-D700E radios has helped with trackers by having the TNC built inside the radio. This keeps the bulk, weight and cost of a tracker down. Be under no illusion, these are incredibly capable radios and well worth the money. There are two basic types of tracker – 'dumb' and 'smart'. A dumb tracker is one that does nothing but send its position report, whereas a smart tracker will have the ability to send and receive messages etc, just as a full-blown APRS station can.

Tracking assets can be invaluable; anything can have a tracker strapped to it, from an ambulance/fire engine/police car through a car/bus/train, to a search and rescue helicopter or search team.

The rest of the world

When Bob Bruninga devised the APRS system, his intention was "to make it a real-time local (tactical) communications system that could be used anywhere, at any time, with no prior knowledge, and would allow everyone in an area to communicate. But this emergency-response design objective naturally is only needed less than 1% of the time.

"As amateurs experimented with it and began putting up fixed networks for routine operations the other 99% of the time, it became obvious that it had the potential for a great wide-area (low-bandwidth) global system." And so the worldwide APRS network began. Here's how.

IGATE (Internet GATEway)

As pockets of activity built up, vast distances needed to be covered to keep the real-time aspect of APRS alive. Realistically this is impossible on RF, so a system called APRServe was developed using the Internet; it is a computer that deals purely with APRS Internet information. IGATEs are APRS stations that are connected to APRServe – they take all transmissions from RF and pass them to the server. This, in turn, sends them to all IGATEs that are connected.

This has given APRS some serious potential, both good and bad. At the moment there are normally in the region of 1500 to 2000 stations being processed by APRServe, and all of their position reports, messages and objects can be sent from the Internet to RF. Obviously, that would be catastrophic and so IGATE sysops have to regulate what, if anything, they send to RF. This is generally a couple of 'exotic' positions intended to promote the IGATE's existence.

What it has done is to enable anyone within range of an IGATE to communicate with anyone else in the world who is also in range of an one.

Remote area operation

It is easy to talk of accessing the worldwide APRS network but what if there is no network where you operate? There have been fairly extensive tests using the satellite 50-35 as a digipeater, all of which have proved that it is possible to access the satellite using a dual-band handheld with a half-wave vertical antenna provided that such operation is not attempted from the centres of large towns.

If satellite operation is possible all that is then required is a dedicated APRSat and ground stations (position gateways) linking APRSat to APRServe, thus completing the network.

For further information see Appendix 1.

Reference

[1] Edited version from *Repeaters* written by the RSGB Repeater Management Committee. See the *RSGB Yearbook* 2002 edition for full details and Appendix 1 for relevant web sites.

CHAPTER 3

Installing radio equipment in vehicles

Introduction

Installing radio equipment in modern cars can present quite a challenge, mainly because of the lack of space to install the rig, the EMC problems and the difficulties of finding somewhere to fix the antenna. This chapter deals with finding a suitable location for the transceiver, plus EMC considerations, while methods of fixing an antenna to a vehicle are described in Chapters 4 and 5.

Before installing any radio transmitting equipment in a new car, it is important to check its owners' handbook for any advice on installing mobile transmitters. If any such advice given by the manufacturer is not followed, this could invalidate the vehicle warranty in the event of any failure of vehicle electronic equipment caused by high levels of RF.

Make sure that radio installation does not interfere with the operation of the vehicle. One space often overlooked is the deployment area of the air bags! Don't install any equipment where it will interfere with the car's safety features.

The instruction manual provided with an amateur transceiver may contain a section on mobile installation. These instructions should be followed unless they conflict with the vehicle manufacturer's instructions.

Location of the transceiver

Of the many considerations regarding the location of the transceiver the most important is finding a suitable space. It needs to be where all necessary controls are within easy reach of the driver/operator but not in such a way as to distract attention from the road.

When you buy a vehicle it is worthwhile considering the location of a transceiver among the priorities governing the purchase. I spent some time looking at various vehicles that would have enough space to mount a normal-size transceiver and found that most required some modification to the interior. Nevertheless there are some such vehicles around and an example of such an ideal mobile shack is shown in Fig 3.1.

Fig 3.1. The mobile campervan 'shack' of Tony Walpole, VK6QG. This layout is of the driving position with an Icom 730 and a homebrew 200W amplifier. The campervan layout allows easy placement of the radio equipment with the controls accessible from the driving seat. This photo was sent to me, with a QSL card, after a mobile-to-mobile contact on 18MHz

My previous two vehicles had a centre console which contained only the stereo/radio unit. I found that this section could easily be removed and left plenty of space for the old IC-707 transceiver, as shown in Fig 3.2. The stereo/radio unit was then fixed to the top of the HF transceiver (not shown in Fig 3.2) with a length of bungee cord.

With either of these solutions described above, ensure that the transceiver is placed so that it does not get a protracted blast of hot air from the vehicle heating system. With most heating systems the air can be diverted to other parts of the vehicle with the heating controls.

Although they are few and far between, there are some vehicles that have dashboards very suitable for amateur radio equipment installation and an example is shown in Fig 3.3.

There will be situations where the solutions for installing an amateur transceiver so far described are inappropriate for you. For example, if you have a large transceiver, or you want to use the shack transceiver for a temporary mobile rig, the only place to put the radio may be the passenger seat. (This assumes you don't have a passenger, or the passenger is willing to sit in one of the rear seats.)

Although I have seen this solution criticised in some amateur radio literature I personally have no problem with it, provided it does not compromise the safety of the vehicle or the occupants in any way – see the safety information in Chapter 1. To this end I have made a 'seat belt' for my transceiver so that it does not fly around in the event of an unscheduled sudden stop. The 'passenger seat' installation is shown in Fig 3.4.

CHAPTER 3: INSTALLING RADIO EQUIPMENT IN VEHICLES

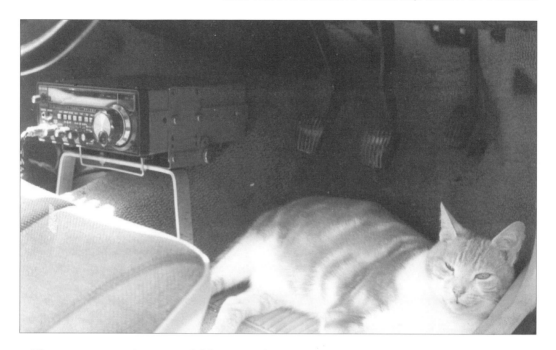

There are transceivers available now whose design helps overcome some of the limitations imposed by vehicles of today. Probably the best known is the Icom IC-706. This transceiver has HF/VHF transmit coverage from 1.6 to 54MHz and 144 to 148MHz, although it transmits only in

Fig 3.2. If you don't have enough room to swing a cat it may be possible to remove the centre full-length console to provide a large space where a transceiver can be conveniently installed and operated. The mounting bracket is fitted with aluminium legs so that it straddles the transmission bump

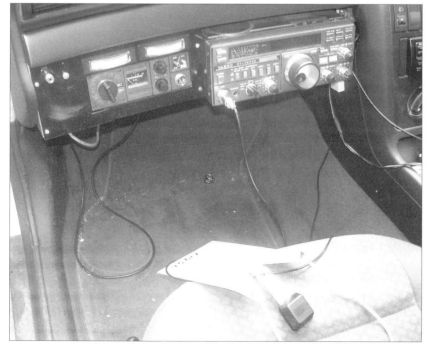

Fig 3.3. The mobile station of G3ROO, with all the equipment mounted in the dashboard. To the right is the FT-747 HF transceiver and to the left an Icom IC-240. HF power and SWR is monitored by the two meters above the VHF radio

23

AMATEUR RADIO MOBILE HANDBOOK

Fig 3.4. The 'passenger seat' installation. In spite of this vehicle being large the options for installing a larger amateur radio transceiver are limited. Removing the centre full-length console is an inconvenient option because of its complexity, so the transceiver is strapped to the passenger seat that has been moved to its most forward position. The IC-737A has ergonomic controls and a large illuminated dial – important features in a mobile rig. It also has the memories and split frequency facilities required for serious DX operating. Note the steering wheel logbook as described in Chapter 2

the amateur radio bands. In addition it has a general-coverage receiver, so it will be able to receive long-wave, AM broadcast band, short-wave broadcast, all amateur bands in all modes from 160m to 2m, FM broadcast band (wide FM), aircraft (118 to 136 MHz AM), VHF from 30MHz to about 160MHz – a versatile little rig to take on a long mobile journey. It is very much easier to find a convenient mounting place for the detachable face plate than a complete radio – see Fig 3.5.

The IC-706 has an innovative, easy-to-operate menu system. It becomes quite easy and natural to use after reading the well-written manual for about an hour; although I don't find it as easy to operate (on the move) as the 'passenger seat' installation shown in Fig 3.4.

Connection to 12V supply

The 12V supply to the transceiver should be fed via positive and negative cables connected directly to the battery terminals with tags or clamps. The cigar lighter socket is a poor source of 12V DC supply because of the poor-quality contacts made by the plug and the socket. Additionally the current rating of this socket is insufficient for use as a power source unless you are using a QRP rig.

Both the positive and negative cables should be fitted with a fuse as close as possible to the battery. The fuse rating should be as recommended by the transceiver manufacturer. The fuses at the battery are not just to protect the radio; they also protect the car. If a short-circuit were to develop in the wiring, the radio fuse would not blow and the wiring could

CHAPTER 3: INSTALLING RADIO EQUIPMENT IN VEHICLES

Fig 3.5. Installation using the IC-706. It is very much easier to find a convenient mounting place for the detachable face plate than a complete radio and in this case the screen forms a sort of 'head-up display'. Although the headset can be plugged into the face plate I chose to plug it into the alternative socket on the radio unit itself, which is mounted under the driver's seat

start a fire. Fusing the radio's positive lead at the battery will prevent this.

The reason for fitting a fuse to the negative cable is to remove a possible safety hazard. If the return current connection for the starter motor developed a high-resistance contact then a high current could flow via the transceiver negative cable and the braid of the antenna feeder.

Routing of DC supply cables

The transceiver DC supply cables should be routed so that they avoid sharp edges or abrasions. Wherever a cable is routed through a bulkhead, a grommet must be used. Sometimes you can find a hole in the bulkhead that is blanked off with a rubber plug. I have made a hole in one of these plugs for supply cables etc – they make a good grommet.

The supply cables should be routed clear of vehicle wiring looms or electrical/electronic units, particularly the engine control unit (ECU) and ignition coil. This is to minimise possible interference to or from vehicle electrical or electronic systems. The positive and negative supply cables may be twisted together along their length in order to reduce noise and interference induced by other wiring.

If it is necessary to take the supply cables from one side of the engine compartment to the other then route them along the side and front as shown in Fig 3.6. It goes without saying that supply cables must be kept

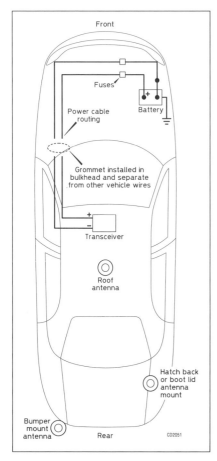

Fig 3.6. Electrical connections from a battery in the engine compartment to a transceiver located by the driver. The electrical connections to the radio are shown routed around the periphery of the engine compartment to place them as far away from other wiring as possible. The ideal location for the battery would be on the left-hand side against the firewall bulkhead. The coax cable runs to the antenna are not shown but are discussed in Chapter 5

clear of fuel pipes, brake pipes, hot components or moving parts. It is a good idea (if possible) to trim the supply cable lengths so that they cannot be accidentally reconnected the wrong way round.

EMC

'EMC' is the general term used to describe mutual electromagnetic interference between the radio equipment and the vehicle electronic system. There are two areas to consider in relation to EMC:

1. Interference to reception caused by electromagnetic energy generated in the engine and ancillaries.
2. Interference by radio transmitters to electronic systems controlling the engine and ancillaries.

Interference to reception

Ignition interference

In recent years interference to mobile amateur equipment has become less of a problem. This is because most modern vehicles have a broadcast radio fitted as standard, so that at least basic interference suppression is included.

Another factor is that most amateur mobile communication now takes place on VHF FM, which is inherently less susceptible to impulsive interference, while impulse limiters on modern transceivers are quite effective in eliminating impulse noise when receiving SSB. Nevertheless, the most common cause of interference to a radio transceiver is the vehicle ignition system.

The noise limiter or blanker operates by using the very short impulses of interference to operate a gate in the receiver IF, and its performance is dependent on a number of factors, including the ability of the receiver front-end to handle large impulses. However, at high engine speeds a noisy ignition system will tend to mask weak signals, even though, with the limiter working, ignition 'pops' may appear to be completely eliminated.

The reason is that strong ignition pulses contain enough energy, when integrated, to block the AGC circuit of the receiver, causing the gain to drop whenever the engine is speeded up. Since the AGC circuits of the receiver obtain no benefit from a noise clipper, it is important that ignition noise be reduced enough at the source that the AGC circuits will not be affected even when the engine is running at high speed.

Reducing ignition interference at source used to consist of either installing a spark-plug suppressor on each plug, or substituting the standard spark plugs with resistor plugs. You could even go as far as completely screening the ignition wiring, as is done on aircraft engines. If you have a very old vehicle then these measures will help reduce ignition interference.

With modern vehicles using engine control units the ignition system is best left alone, except for ensuring that it is well maintained, such as replacing old high-tension wiring that may have become leaky, ensuring the rotor and distributor cap are in good shape and that the spark plugs are clean and the gap settings are correct.

Sometimes high-voltage impulses can get into the low-voltage wiring by the wiring layout. It is worth checking to see if any of the high-tension wiring is cabled with low-tension wiring, or run in the same conduit. If so, re-route the low-tension wiring to provide as much separation as practical.

Alternator whine

This type of interference occurs while the engine is running and its pitch varies with engine RPM. I don't know if it is that I have just been lucky but in 40 years of mobile operation I have never experienced this problem.

You can buy alternator suppression capacitors (see Appendix 1) that have special fittings so that they can be fixed to the side of the alternator. The unit I examined at a local car parts store was claimed to have a value of 3μF.

A filter in the power leads may help and is described later.

Electric motor interference

Electric motor noise is a common cause of interference and I have found that windscreen wiper motors are one of the worst offenders. Some motors, such as those that power electric windows, are used only occasionally and can be tolerated. Others, such as windscreen wipers, are a different matter. You can buy (see Appendix 1) interference suppressors which comprise an in-line choke, usually with a current rating of up to 7.5A – these can be fitted in the motor leads of windscreen wipers, heater blowers or fuel pumps.

Miscellaneous

The following are a collection of causes of interference which I have read about, although not actually experienced. One of these is electronic engine control unit radiation, which sounds like the digital noise you often receive if your radio is near a home computer system or video game.

Loose linkages in body or frame joints anywhere in the car are potential static producers when the car is in motion, particularly over a rough road. Locating the source of such noise is difficult, and the simplest procedure is to give the car a thorough tightening up in the hope that the offending poor contacts will be caught up by the procedure. The use of braided bonding straps between the various sections of the body of the car also may prove helpful.

In many cases the control rods, speedometer cable etc will pick up high-tension noise under the bonnet and conduct it up under the dashboard where it causes trouble. If so, all control rods and cables should be bonded to the firewall (bulkhead) where they pass through, using a short piece of heavy flexible braid of the type used for shielding.

In some cases it may be necessary to bond the engine to the frame at each rubber engine mount in a similar manner. If a rear-mounted whip is employed, the exhaust tail pipe should also be bonded to the frame if supported by rubber mounts. Check to see that the bonnet makes good ground contact to the car body at several points.

At high speeds, under certain atmospheric conditions, corona static may be encountered unless means are taken to prevent it. Broadcast receiver whips often use a plastic ball tip to minimise this type of noise, which is simply a discharge of the frictional static built up on the car. A whip that ends in a relatively sharp metal point makes an ideal discharge point for the static charge. This will cause corona, and associated interference, at a much lower voltage than if the tip were hooded with insulation. A piece of shrink-wrap sleeving slipped over the top portion of the whip and heated will prevent this type of static discharge under practically all conditions. An alternative arrangement is to wrap the top portion of the whip with a couple of layers of insulating tape.

Electromagnetic interference to vehicle electronics

RF interference to electronic control systems has become a factor in mobile radio, not least because the possible safety aspects. Basically the problem is caused by the possibility of RF energy being picked up in the wiring of the car and entering the logic of the control circuits. The effects can vary from specific faults, such as failure of door locks or erratic flashers, to complete failure of the engine control system.

If you are going to install an amateur radio transceiver in a vehicle that you already own which has an electronic engine control unit, then it is probably a good idea to commence with a temporary installation. Use a separate battery large enough to allow the transceiver to produce close on full power output when transmitting. Otherwise, run fused power leads from the passenger compartment through an open window and under the bonnet direct to the battery, making sure that these power leads are located as far from any other cabling as possible. The transceiver can be placed on the seat as shown in Fig 3.4. The antenna can be fixed to a magmount or attached to a towbar if you have one.

With the vehicle stationary and the engine running the following should be checked while the transmitter is being operated at its maximum power:

- There is no apparent engine misfiring.
- No warning lights flicker or come on.
- The direction indicators flash at the normal rate.
- The windscreen wipers operate normally.
- There are no unwanted effects on other electronic systems, such as central locking or air bags.

If any unwanted effects occur, it will be necessary to relocate the antenna, reduce transmitter power or both. It should be noted that not all possible adverse effects can be detected when the vehicle is stationary, for example, anti-lock braking, cruise control, automatic transmission, electric power-assisted steering. It is therefore advisable to test drive the vehicle in a suitable location, preferably off the public highway. If any effect such as engine misfiring is noted when the transmitter is operated, transmission should cease immediately.

I have heard of one situation where the coaxial cable was routed close to the power leads of the radio. Whenever the transmitter was keyed the engine started missing. Moving the coax away from the power leads eliminated the problem.

Now while all this may sound rather daunting I should point out that there are many HF and VHF mobilers on the road who experience no adverse affects to the vehicle whatsoever. Vehicle electronic systems must have some degree of EMC protection – a vehicle that developed a failure every time it was near a high-power broadcast station would hardly provide good publicity. G3YXM uses 400W with a homebrew linear and the worst effect that he has experienced was the embarrassing triggering of the vehicle alarm while driving.

Filtering

Power lead filters have been found to be effective in many different cases of EMC and may be effective in reducing electrical noise on receive or interference to the vehicle electronics.

Filtering may be simply the action of installing snap-on ferrite cores around the power leads where they entered the engine compartment. While this may help reduce the electromagnetic noise that reaches the radio, the main reason is to prevent stray RF from getting onto the power cables and being carried back to the engine computer. (More than one snap-on core may be required for HF applications.)

G3MEW cured a problem of alternator whine on his FT290 (144MHz) after finding that simply adding a suppressor to the alternator made no difference. The supply lead filter comprised a single layer winding around a ferrite ring core with a 2200pF capacitor to earth on the rig side of the winding.

A more elaborate filter can made using two FT-140-43 ferrite cores with about 10 turns of each lead wound onto the core. You can also try installing 10000nF capacitors from the positive lead to the negative lead, or from the positive end to chassis ground (or both). The capacitors may work best with or without one or both of the ferrites.

Commercial filters are now available for mobile power leads and an example is shown in Fig 3.7.

Further reading

MPT 1362 (1997 edition), *Code of Practice for installation of mobile radio equipment in land based vehicles*. Available from the Radiocommunications Agency of the DTI.

Fig 3.7. The Icom OPC-639 power lead filter, designed specifically to reduce EMC problems in mobile installations. It comprises two ferrite cores and four surface-mount capacitors. Both supply leads are wound on the ferrite cores and the capacitors connect from the leads to ground, between the two ferrite cores. For availability see Appendix 1

Radio Telephone/Mobile Radio Installation Guidelines issued by General Motors Corp (USA).

RSGB EMC Committee *Leaflet EMC 06 Automotive EMC for Radio Amateurs.*

CHAPTER 4

Mobile antennas

This chapter is devoted mainly to HF antennas because HF poses the greatest challenge to mobile antenna design and installation; however, VHF antennas are not overlooked and are described later in the chapter.

The antenna is the key to successful mobile operation. The vertical whip antenna is the most popular mobile antenna, regardless of the band in use, due to the shape of the vehicle, space limitations and the slipstream caused by the vehicle motion. The easiest way to feed such an antenna is to make it a quarter-wavelength long at the frequency in use. The resonant quarter-wavelength is a function of frequency and is 2.5m (8ft 2in) on 28.4MHz, 1.48m (58.5in) on 50MHz and progressively shorter on the higher VHF bands. Quarter-wave antennas on the 28MHz bands and higher are quite practical but on the lower HF bands it is a different matter. Even on 21MHz a quarter wavelength is 3.45m (11ft 2in) and on 14MHz is 4.99m (16ft 4in).

Inductive loading using a coil

It follows that a practical antenna for the HF bands will be shorter than a quarter-wave long. For a given antenna length, the feedpoint exhibits a decreasing resistance in series with an increasing capacitive reactance as the frequency of operation is lowered. In order to feed power to such an antenna it must be brought to resonance so that the feedpoint is resistive. This is achieved by adding some inductance and is known as *inductive loading*.

By far the simplest method of achieving resonance is to use a variable inductance mounted directly below the base of the antenna. This is the method used on military and marine installations where the main consideration is for a rugged antenna that does not interfere with operational considerations. It has the advantage that the inductance can be varied from within the vehicle. The main disadvantage of such an arrangement is that when the antenna is electrically short the efficiency is low. At this point it may be of use to consider why this should be.

For antennas that are a quarter of a wavelength long or less, radiation is proportional to current flow and the antenna length. However, current

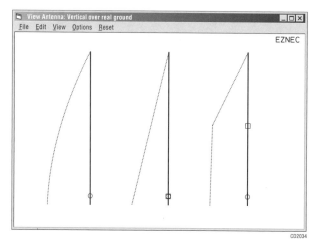

Fig 4.1. A computer model of antenna current in vertical antennas. The circle represents the feedpoint; the square represents the loading coil. (a) The current distribution of a full-size quarter-wave vertical antenna. (b) An electrically short vertical with a loading coil at the base of the antenna. (c) Short vertical antenna with the coil in the centre

distribution is not uniform. For example, current distribution of a full-size quarter-wave vertical antenna varies with the cosine of the length in electrical degrees (90°) and is illustrated using a computer model of antenna current shown in Fig 4.1(a). For a short antenna (say $1/8$ wavelength) resonated with a base loading coil, the current distribution decreases linearly from the base to the top of the antenna as shown in Fig 4.1(b). The efficiency of a short antenna can be improved by placing the loading coil at its centre.

We can get some idea of the power density of the signal radiated from these antennas using a method described by Les Moxon [1]. Here the current distributions are drawn on graph paper as shown in Fig 4.2. The relative power density of the radiated signal can then be estimated by counting the number of squares under the curve. From this it can be seen that most of the radiation is from the lower section of the antenna. By placing the loading coil in the centre of a short antenna, the lower section, carrying the greater current, is extended.

From Fig 4.2 it would appear that more the loading coil is elevated, the greater will be the efficiency of a loaded antenna; however, this does not tell the whole story. As the loading inductance is elevated, the inductance value has to be increased. This is because the capacitance of the shorter antenna section (above the coil) to ground is now lower (higher capacitive reactance), requiring more inductance to tune the antenna to resonance. A larger inductance means more turns on the loading coil and greater coil resistance losses.

This loss of capacitance on an antenna with an elevated loading coil can be offset by using a *capacity hat*; although this improves antenna efficiency it makes for a more complex structure. For mobile use centre loading has been generally accepted as the best compromise between efficiency and construction limitations.

The time-honoured method of multibanding such antennas is to have a separate coil for each band which is removed and replaced with another to change bands; this is the method used by G3TSO [2] and G3MPO [3] in their constructional designs of centre-loaded mobile antennas.

Such complexity may not be necessary; after all, ATUs and the old transmitter pi-output tuned circuits use tapped coils, where the unwanted inductance is shorted out.

An example of such an arrangement is the commercial Texas Bugcatcher mobile antenna shown in Fig 4.3. Here adjustable taps are set experimentally then the band can be changed simply by clipping a fly-lead on to the appropriate tap.

CHAPTER 4: MOBILE ANTENNAS

Coil construction

The construction of a loading coil for a mobile antenna must be rugged to stand up to weather and the mechanical strain of a fast-moving vehicle slipstream. The following represents a couple of solutions.

The G3MPO coil and antenna

This design uses a single antenna structure with a different coil for each band. This coil was produced using workshop facilities consisting of little more than a Workmate®, electric drill and taper taps and dies. According to G3MPO, this design has proved completely secure in several thousand miles of motoring. Full use was made of a local plumber's stockist's supply of ready-made brass, stainless steel and plastic bits and pieces. 15mm plumber's brass compression couplers were selected as both coil terminations and the means of fixing them into the antenna.

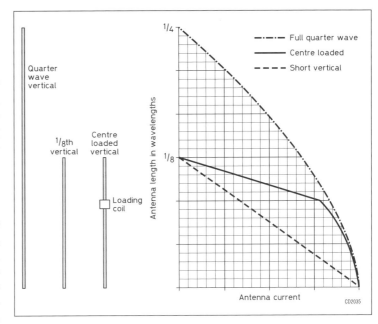

Fig 4.2. Current distribution, from Fig 4.1, drawn on graph paper. Relative radiation power densities of the antennas can be estimated by counting the number of small squares under the appropriate curve

Because the fittings and the antenna material are an integral part of the design these are described as well as the coil. The bottom section of the antenna is made from a length of 15mm stainless steel central heating tubing.

White polypropylene waste pipe proved a good choice for a coil former. The thread of the brass 15mm compression couplers can be screwed (with some difficulty) into the end of the 20mm (0.75in) version of this tubing to make a very strong joint. The ends of the tube can be pre-heated in hot water if necessary. Even better, a 12.5mm (0.5in) BSP taper tap can be

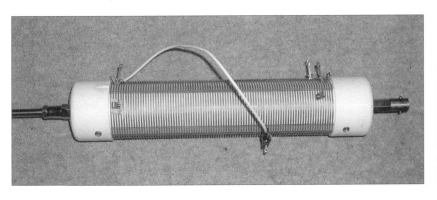

Fig 4.3. Coil section of the Texas Bugcatcher mobile antenna showing the preset tapping points. The appropriate unwanted section of the coil is shorted out using a fly-lead with a crocodile clip

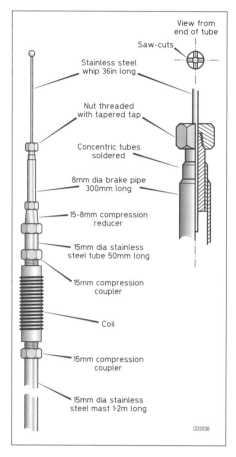

Fig 4.4. (a) Section of the G3MPO mobile antenna. (b) Detail of stainless steel whip/ 8mm diameter brake pipe clamp

used to cut a starting thread in the tubing. A second coupler screwed into the other end gives an excellent coil former with ready-made 15mm connections at each end, which fit and clamp directly onto the 15mm diameter lower mast section.

Varying lengths of former are used for the higher-frequency coils and, where a greater diameter is needed for the lower frequencies, the same 20mm (0.75in) former is used as a spine. This runs up the middle of a larger-diameter tube to which it is attached by packing the space between them at each end with postage-stamp-size pieces of car-repair glass mat soaked in resin. The coil assembly can then be waterproofed with a silicone rubber sealant.

The whip above the coil former comprises a short length of small-diameter tube, fixed to the top of the coil with a 15mm coil coupler. A length of stainless steel whip is slid inside this tubing. Suitable tubing can be obtained from a car accessories shop in the form of copper brake tubing, which comes in outside-diameter sizes from 8mm down. A 230mm (9in) length of 8mm diameter pipe, with one end plugged by short lengths of the next two sizes down and soldered into position, gave a nice sliding fit to the 2.5mm diameter whip.

Some sort of quick-release lock is required to hold the whip in position once its length is set. This is achieved by cutting a thread on the last 12mm of the end piece of tubing with a thread-cutting die and making cross cuts down its length with a mini-hacksaw.

A matching nut with a tapered thread can be made from a short length of brass rod, drilled and taper-tapped so that it would close the tube down on to the end whip as it is screwed on, thus locking it. The whip structure is completed by connecting the telescopic section onto the coil using a 15mm-to-8mm (microbore) brass reducer and a 51mm (2in) length of 15mm tubing, as shown in Figs 4.4 and 4.5.

The best method of attaching the wire to the end couplers is to drill two small holes through the polypropylene just beyond where the end of the coil would lie, and pass the wire into the tube and out through the coupler to which it is then connected. It was found best to solder a hairpin of wire onto the inside of the coupler before fitting it into the plastic former. The coil wire was then easily soldered onto this pigtail at the appropriate time.

Two or three lengths of double-sided tape are then stuck to the coil former. A sufficient length of enamelled copper wire is cut for the coil in question. Seven times the number of turns, multiplied by the diameter of former (from Table 4.1), allows enough to wind the coil with some to spare. The wire spacing is achieved by winding two lengths of wire onto the former side by side and subsequently removing one of them. The double-sided tape fixed to the former holds the remaining winding in position. 10 to 12mm of wire is wound beyond the holes through which the wire endings were taken and, after removing the spacing wire, the winding is coated with polyurethane varnish. When dry, the coil is wound back at each end until the required number of turns are obtained, and the ends fed through into the former, out through the end couplings and soldered to the coupling hairpins. The two small holes in the former can be sealed with varnish or mastic and the winding bound with a double layer of self-amalgamating tape.

Soldered connections are pushed well down into the coupling, out of the way, and the coil given two coats of polyurethane varnish. The self-amalgamating tape can be omitted if you prefer the appearance of varnished copper coils, but it is easy to use and provides additional protection against knocks and bangs.

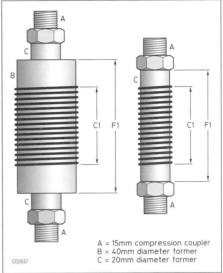

Fig 4.5. Coil former construction and dimensions of G3MPO antenna. *Fl* is the coil former length and *Cl* the coil winding length – see Table 4.1

The G3LDO loading coil

This loading coil arrangement has been in use for about 12 years and has given good service. It is designed for frequencies between 14 and 29MHz and uses a large-diameter, self-supporting coil. It can be designed as a multiband antenna by simply shorting out the unused section of inductance. Unlike other designs no attempt is made to protect the coil from the weather except to spray the whole structure with spray grease to prevent corrosion. I have devised two methods of overcoming the insulator

Table 4.1. Coil data – read in conjunction with Fig 4.5							
F (MHz)	D (mm)	Fl (mm)	Wire (SWG)	N	Cl (mm)	L (μH)	Rr (Ω)
29.0	20	77	18	9	28	0.9	35
24.9	20	90	18	15	44	1.7	29
21.2	20	115	18	23	64	3.0	22
18.1	20	140	18	34	92	4.5	17
14.25	20	146	20	45	96	8.4	11
10.13	40	115	20	31	72	19	6
7.05	40	165	20	58	117	41	3
3.65	40	305	22	130	157	153	0.8
1.9	40	280	28	294	236?	558	0.2

F is the frequency (MHz), *D* is coil former diameter (mm), *Fl* is the length of coil former tube (mm), *N* is the number of turns, *Cl* is the length of coil winding (mm), *L* is the coil inductance (microhenrys), *Rr* is the theoretical radiation resistance (Ω).

AMATEUR RADIO MOBILE HANDBOOK

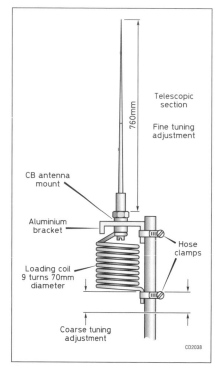

Fig 4.6. The insulator in the vertical section of the antenna for the loading coil is provided using an antenna base fitting mounted on a metal bracket. This design is for 14MHz although it can be used for higher frequencies by shorting out turns on the coil

design with minimal workshop tools. Both make use of the antenna constructor's friend – the hose clamp or jubilee clip.

The first design uses a conventional mobile ham radio or CB antenna base as the insulator. The loading coil is made of thick copper wire and constructed as shown in Fig 4.6, with one end connected to the antenna base and the other to the lower vertical section of the antenna by a hose clamp. The coil can be made self-supporting by using thick wire for the inductor and quite a wide range of tuning can be obtained by altering the position of the hose clamp. The top section was made from the base section of a military vehicle antenna, the screw fitted to the CB antenna fitting. On top of this was soldered a piece of telescoping antenna, as used in portable transistor radios, for fine-tuning the antenna.

When transmitting there is a high voltage across this antenna mount and any moisture inside the antenna base insulator will cause trouble. I found the best solution is to coat the insulator with grease.

In the second design the insulator is made from a short section of white polypropylene waste pipe whose internal diameter is approximately equal to the aluminium tubing elements, and held together with hose clamps – see Fig 4.7. The coil is made from 2mm diameter wire held in place with strips of Perspex with holes drilled at regular intervals to produce a self-supporting inductor. (You could also use B&W coil stock Part 3033 – see Appendix 1 for availability.) The coil is held in place with additional hose clamps; the top one also clamps the telescopic top section (and capacity wires if necessary). A lead with a crocodile clip termination is used to short out sections of the coil for the higher-frequency bands. If you use thick wire for the core then insulating supports are not necessary and the inductance of the coil can be varied by adjusting the position of the lower hose clamp.

The insulator and the hose clamps can be coated with a thin layer of grease, which provides a waterproof coating and prevents corrosion. This grease picks up some road dirt after several months driving but it is easy to remove it with a cloth and then renew it.

The continuously loaded HF antenna

Another common form of mobile antenna is the helical whip. This type of antenna comprises a fibreglass whip with the loading coil wound along its whole length, a technique known as *continuous loading*. This construction does away with the mechanical construction problems which may be encountered with a loading coil design. Fibreglass whips are cheap, light, and exhibit much less wind loading than centre-loaded vertical antennas.

A description of how this type of antenna radiates is given by G6XN [1] and is illustrated in Fig 4.8. The horizontal components cancel, the vertical ones add to produce the equivalent of a current I flowing through a distance d equal to the pitch of the turn on the right. The losses are increased compared with those in a straight conductor of the same wire diameter in the ratio $\pi D/d$ or rather more unless the turns are spaced by several times the wire diameter.

Design data for a 145MHz version was described by G8ENN [4]. Some HF design data for a linear pitch helix was included in reference [4] and is given in Table 4.2.

Fig 4.7. Insulator made from a short section of white polypropylene waste pipe whose internal diameter is approximately equal to the aluminium tubing elements and held together with hose clamps. The coil, telescopic top section and capacity wires are also held with additional hose clamps. The coil was designed for the 14MHz band although any number of turns can be shorted out for the higher-frequency bands

G6XN suggested that a 3.5MHz mobile helix might consist of a quarter-wavelength of 14SWG (2mm) wire wound on a 2.4m (8ft) fibreglass rod, 22mm (1in) in diameter. However, the *ARRL Antenna Book* [5] says that (from experience) a section of wire approximately one half-wavelength long with a linear pitch will come close to yielding a resonant quarter-wavelength. This is in agreement with the data by G8ENN shown in Table 4.2.

The construction details of the commercial helix antennas in my possession are as follows.

The bottom section (28MHz) of a G-Whip, see Fig 4.9, has 280 turns wound on a fibreglass whip, 1m long. This whip tapers from 12mm to 9mm so the wire length of 8.8m is more than a half-wavelength. The coil section of the Q-TEK five-band HF mobile antenna, see Fig 4.9, has 133 turns wound on a 13mm fibreglass whip, 1m long. This antenna uses 6.7m of wire, again longer than one half-wavelength on 28MHz. Most commercial designs use non-linear pitch winding, ie wider spacing (and sometimes thicker wire) for the lower portion of the antenna.

Equally, there are several approaches to multibanding. With the ProAm antenna (two centre antennas in Fig 4.9) the solution is to have a completely separate antenna for each band. The G-Whip for the 28, 21 and 14MHz bands (shown to the right of Fig 4.9) comes in three sections. The 28MHz base section can be extended with plug-in 21MHz and 14MHz sections. The lowest cost (and in my opinion the best cheap technical solution to the multiband problem) is the Q-TEK five-band HF mobile

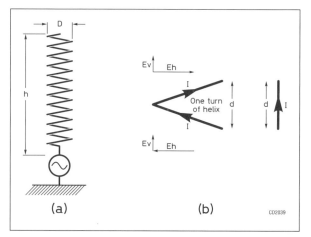

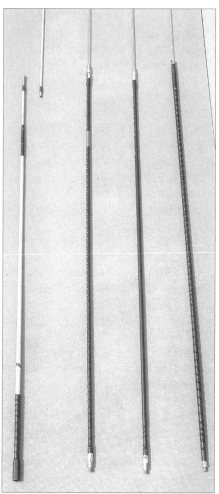

Above: **Fig 4.8.** Helical whip, where (a) shows schematically a vertical helix and (b) is an enlarged view of one turn, showing the horizontal and vertical fields produced by each half-turn as seen by a distant observer

Right: **Fig 4.9.** Four types of commercial helical mobile antennas. From right to left: (a) three-section G-Whip for the 28, 21 and 14MHz bands with the 21MHz section plugged into the lower 28MHz section and the 14MHz section shown separately. (b) 14MHz ProAm antenna. (c) 80m ProAm antenna. (d) Q-TEK five-band antenna

antenna, which uses a basic tapered coil section with a selection of plug-in capacitor top sections. Other solutions to multibanding are described later but they are not cheap.

The 27MHz CB era brought about some interesting helix mobile antenna designs. I have two such antennas, see Fig 4.10, that have the brand name 'Firestick'. The largest is 2m long wound with 5.4m of wire. It has a very pronounced non-linear pitch winding with one turn per 60mm at the base and close wound at

Table 4.2. Data for a 145MHz helical whip by G8ENN

Band (MHz)	Number of turns N/height h in metres (ft)				
	0.61(2)	0.91(3)	1.25(4)	1.52(5)	1.83(6)
1.8		3072	3254	3400	3526
3.5	1457	1579	1671	1746	
3.8	1341	1454	1539	1608	
7.05	722	782	826	869	
14.15	358	386	405	418	
21.2	238	254	263	266	
28	178	188	191	187	
30	166	174	175	169	

Number of turns (N) for helical whips of various lengths (h) for the HF bands. This data is for whip formers with an average diameter of 12mm (0.5in). Note that G8ENN made this data available before the WARC bands were available but approximate data can be derived by interpolation. On at least the higher-frequency HF bands a bifilar helix should be reasonably easy to wind and would quadruple the feed impedance of the antenna.

the top. The antenna had a reputation among the CB fraternity for being very efficient; rumour has it that it was banned as a CB antenna. I have removed around 8 turns from the top of the antenna to bring it into resonance in the 28MHz band; I have found it to be an excellent antenna for mobile DX.

The second antenna is 1.4m long; wound with 190 turns (very close to the 28MHz design for a 1.8m (6ft) long antenna shown in Table 4.2), using a similar pronounced non-linear pitch winding as the first antenna (one turn per 50mm at the base and close wound at the top). Around four turns are removed from the top. A telescopic antenna from an old transistor radio is fixed to the exposed section of fibreglass using a small section of metal tubing and a hose clamp, and the end of the helix soldered to it. By altering the length of the telescopic section the antenna can be tuned to any frequency between 14 to 25MHz.

G3YXM constructed a high-performance helical HF whip to cover 20m to 10m. Multibanding is achieved in the same manner as the Outbacker commercial antenna, with jumper leads to short out sections of the coil. The basic whip has a coil wound with 5mm ($^3/_{16}$in) copper strip at the top of a 1.5m (5ft) fibreglass tube. A stainless steel whip section is fixed to the top. The basic whip resonates on 18MHz and is brought down to 14MHz by connecting a small coil at the base of the antenna. For 21MHz some of the coil is shorted and for 28MHz it is shorted out altogether.

The availability of commercial helix antennas will change due to manufacturing trends and stockists' buying practices. At the time of writing Watson and MFJ are two of the most popular of these types of antenna on the market here in the UK. However, close inspection of the construction will show that they are similar to those already described; this is dictated by mobile antenna size constraints and the laws of physics.

The W6AAQ continuous-coverage HF mobile antenna

Back in the early 'sixties a commercial mobile antenna was marketed under the name 'Webster Band Spanner'. The coil was wound on the upper inside portion of a fibreglass tube. The top section of the antenna comprised a stainless steel plunger with contacts to short out the unused sections of the coil. This allowed the antenna to be resonated on any frequency between 3 and 30MHz and then locked using a compression lock. The construction of this antenna is shown in Fig 4.11.

More recently (1992) I received details from Don Johnson, W6AAQ, of a similar design to the Webster Band Spanner. In this design the antenna resonance is adjusted from the driver's/operator's position via a remote-controlled cordless screwdriver electric motor. The construction of the antenna is different with the coil mounted on a lead screw inside a 1m (3ft) long 50mm diameter aluminium, brass or copper tube as shown in

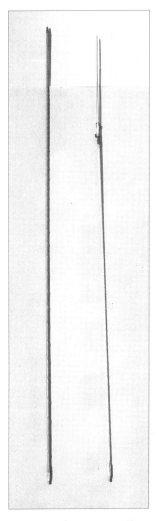

Fig 4.10. Two Firestick CB antennas modified for the amateur bands. The one on the left is modified for 28MHz by removing around eight turns from the top. The smaller one has around four turns removed from the top. The end of the wire is soldered to a telescopic antenna from an old transistor radio. The arrangement allows the antenna to be tuned to any frequency between 14 to 25MHz

Fig 4.11. Advertisement for the Webster Band Spanner from June 1962 *QTC* (Journal of the Radio Society of East Africa). Because of the ability of this antenna to be resonated on any part of the HF spectrum it was also used by the numerous government and commercial HF networks

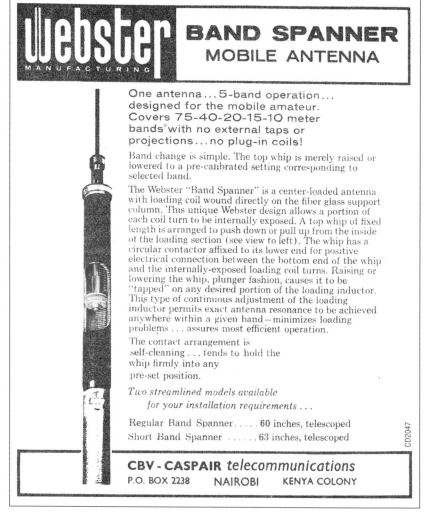

Fig 4.12. As the motor is rotated the coil is raised or lowered so that more or less of the coil is contained within the lower tube section. Finger stock connectors are used to short the coil to the tube to obtain the appropriate resonance.

A circuit of the antenna and the control box is shown in Fig 4.13.

The DK3 is also marketed by N7LYY. There is also the HF-1500 from High Sierra and the WBB-3 version made by Knott Ltd. This antenna is available in the UK (see Appendix 1 for availability).

Regarding the WBB-3, Peter Waters, G3OJV, notes [6] that he has tried every conceivable mobile antenna and is convinced that the larger-diameter coil designs offer much better efficiency, with tests giving up to 6dB of gain on 80m from a 2in diameter centre-loaded whip compared to thinner models. Last year his favourite antenna was the Bugcatcher described earlier. However, it meant getting out of the vehicle to switch

CHAPTER 4: MOBILE ANTENNAS

bands and, when it rained, the open coil suffered a slight drop in frequency on 80m and 40m because of the presence of moisture. However, it was the best antenna he had tried and he still regards it as a benchmark mobile antenna.

The WBB-3 antenna is not lightweight, tipping the scales at around 4.5kg, so don't even consider using a magnetic mount! It is designed to mount on a flat surface through a ³/₈in hole. The necessary ³/₈in bolt, washers, insulators etc are supplied. G3OJV uses a metal bracket made up by the local body repair shop that is attached to the underside of his vehicle. It wasn't expensive and made a very rigid base on which is mounted a ³/₈in ball mount so that a wide range of HF antennas could be tested. The antenna installation is shown in Fig 4.14.

The tuning of the antenna is achieved via a control box, which has a large tuning button and a toggle switch that selects the direction of travel. 4m of control cable are supplied and the necessary water-resistant plug and socket is provided to connect the control cable to the base of the antenna. A shorter cable also emerges from the box that needs to be connected to the car 13.8V supply. This can be achieved by simply attaching a cigar lighter plug to the cable. Mounting the antenna through the ³/₈in hole is very

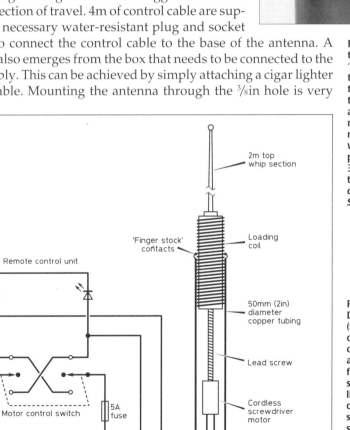

Fig 4.12. Detail of the tuning section of the 'screwdriver' antenna, which shows the coil and fingers that short the turns as it emerges. This is normally hidden beneath the white weather shield. This photo is of the WBB-3 derivative – see text. *(Photo courtesy of Waters and Stanton)*

Fig 4.13. W6AAQ's DK3 mobile antenna (not to scale). The control box is located by the driver and power obtained from either the rig supply or the cigar lighter socket. The original had relay switched capacitors, selected from the driver's control box, to match the antenna on the lower frequencies

41

AMATEUR RADIO MOBILE HANDBOOK

Fig 4.14. The WBB-3 screwdriver antenna as installed by Peter Waters, G3OJV, on the rear of his vehicle. (*Photo courtesy of Waters and Stanton*)

simple and takes only a couple of minutes. Likewise, removing it from the vehicle takes about the same length of time.

To tune the antenna you need to be able to measure the VSWR. G3OJV uses an IC-706 transceiver and the only way to generate a continuous carrier is to switch to AM or FM. However, there is a device, made in the USA, which plugs into an accessory socket on the back of the IC-706 and allows you to transmit around 10W of RF just by pressing the tune button on the transceiver. It is a very convenient way of generating a carrier for tuning and times out after around 20 seconds. It was then a simple task of pressing 'Tune' on the IC-706 and then holding down the button on the antenna control unit (having first selected direction of travel – HF or LF) and waiting for the VSWR to come down. As supplied, the antenna coil is fully retracted. Initially the antenna was tested on 40m. It took just over 20 seconds for the VSWR to come down to what the IC-706 saw as 1:1.

The WBB-3 comes supplied with a matching transformer but it was not possible to get a very low SWR on the higher-frequency bands. A capacitor matching box (the MFJ-910) was used in place of the supplied transformer (see 'Matching the antenna to the feeder' later in this chapter). The MFJ-910 uses the principle of adding capacitance across the feedpoint to provide matching. A rotary switch is used to select different values of capacitance. One position switches out all the capacitance when the antenna is used on 14MHz and above.

And how did it perform? G3OJV says "I was now able to tune to any band quickly and achieve a true 1:1 VSWR wherever I wished to operate. It was pure heaven. As I was about to set off into the Highlands of Scotland, it was a great opportunity to assess the performance and no doubt get it thoroughly wet in the process! The weather was not bad but I did experience one day of what can only described as wall-to-wall rain and several heavy showers at other times. In this respect the antenna seemed to be unscathed and there was no noticeable frequency shift. I operated on SSB all bands and the DX has included VK, JA, HS0, Africa plus many USA contacts and of course Europe. I felt the performance was the same as the Bugcatcher, but with greater ease of changing bands. It was a joy to be able to switch bands without having to get out of the car. I can truly say that this has been the best mobile antenna that I have ever used. Although it is not cheap, you do get a lot for your money. Many users will no doubt be happy with the impedance transformer supplied and not be fussy like me and install the MFJ-910. Previously I had used the matching coil auto ATU to allow me to move around the LF bands when the VSWR began to rise. This is not needed now, so there was a big saving in expense here.

The top whip section is not a magical length. It is simply felt to be a practical length and one that is needed if you require 80m operation. If you don't, then just shorten it. Resonance is achieved by adjusting the tuning via the remote control box. Likewise, you can install a longer whip section (³⁄₈in fittings are easy to come by) and thereby achieve improved performance. In fact I tried this with varying whip lengths, and for parking up a really long length could be installed. However, do remember that you will then have problems with 10m and 15m because the antenna will be too long for quarter-wave resonance."

Fig 4.15. This is what the Vaeriel mobile whip looks like on the car. The little guy-lines stop it swinging from side to side or bending back too much at speed

The G3YXM Vaerial antenna

One of the problems of mobile operation on the lower frequencies is the narrow bandwidth of an efficient mobile antenna. Dave Pick, G3YXM, has built an antenna that gives an adjustment range of 50kHz on 160m and 100kHz on 80m from a control box by the driver's seat.

The antenna is just under 3m long and mounted on the roof of the car on a cross member bolted to the luggage rails. (see Fig 4.15).

The component parts of the Vaerial antenna are shown in Fig 4.16. The coil is the second item down, and is wound on a 470mm long piece of 42mm plastic 'solvent weld' pipe (this is stiffer than the normal plastic tubing) with about 304 turns of 1.25mm enamelled wire (for 160m) and covered in heat-shrink sleeving. At the top of the picture is the base section (600mm length below the coil) which is based on a 10mm diameter fibreglass rod, the bottom of which is wound with thick copper wire and covered in heat shrink. The plastic tube containing the motor, the lead-screw and the copper slug is attached using Isopon fibreglass resin car filler. The 150mm length of 12mm copper water pipe is driven up and down inside the tube by the motor and lead-screw (M4 threaded nylon

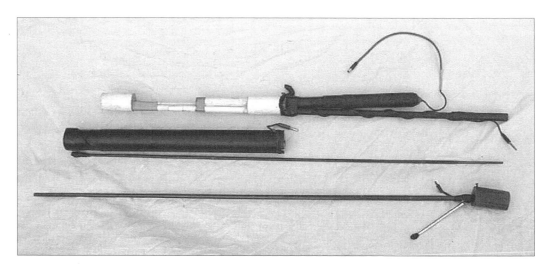

Fig 4.16. The component parts of the Vaerial antenna. The second item down is the coil wound on a 470mm long piece of 42mm plastic tube. The base section (600mm length below the coil) is based on a 10mm diameter fibreglass rod containing the motor (top), the lead-screw and the copper slug. At the bottom of the picture is the 1770mm top whip section

rod) and hence in and out of the coil. The slug and the lead-screw is coated in silicon grease to prevent it the slug rotating in the tube.

The two parts of the ex-army 'tank whip' are copper-plated steel tube sections, total length 1770mm, and the section which fits into the top of the coil has a small telescopic whip sticking out – this allows for adjustment of the centre frequency of the system.

The 80m coil has about 100 turns of 1.5mm wire, close wound on the bottom of another identical piece of tubing – it would be better to wind the coil near the top of the former but then the copper slug wouldn't go inside the winding.

The system will stand the 400W from the MOSFET PA but watch out for sharp edges at the top of the whip. Smooth them off and insulate well with self-amalgamating tape or there will be a firework display if you use the antenna with this power.

Matching the antenna to the feeder

All the antennas so far discussed are fed with 50Ω coaxial cable – normally the centre is connected to the antenna and the braid to the vehicle body. However, the radiation resistance of the antenna will generally be lower than 50Ω and, for a given antenna size, depends on frequency. Typical radiation resistance figures for a 2.4m (8ft) antenna are as follows:

Frequency (MHz)	1.8	3.6	7.05	10	14.2	18	21.3	25	28.5
Radiation resistance (Ω)	0.2	0.8	3	6	12	17	21	28	36

In practice the feed impedance will include the RF resistance of the loading coil and the resistance losses. The loss resistance, taken in total, is usually much greater than the radiation resistance at the lower operating frequencies. For example, the radiation resistance of an 80m antenna is around 1Ω and the loading coil resistance may be around 10Ω. The ground loss will be between 4 and 12Ω, depending on the size of the vehicle, so the feed impedance could be in the region of 20Ω. This will give an SWR of 2.5:1 at resonance, which gets progressively worse very quickly as the transceiver is tuned off the antenna resonance, clearly beyond the impedance range of a modern solid-state transceiver 50Ω PA (unless it has a built-in ATU).

At the other end of the HF spectrum the radiation resistance is much higher and, even though the coil's losses are lower, a transceiver can be connected directly to the antenna via a length of 50Ω coax.

A non-resonant antenna can be used with an automatic ATU provided the antenna and tuner are close together. It is not as efficient as the resonant antenna but performs quite well with a helical antenna that is less than half-wave resonant on the highest frequency. Such an arrangement is uncommon in amateur use and is used mainly where the antenna installation has to be very rugged, such as in military systems. An amateur installation is shown in Fig 4.17.

Fig 4.17. The automatic ATU installation in the boot of G0GCQ's car. The electrical connections to operate the ATU are run directly from the battery and through a local fuse box. The thick wire in the foreground is the antenna wire running to a helix antenna mounted on a hatchback antenna mount

There are several ways of matching the nominal 50Ω transceiver output to the impedance encountered at the base of a resonant mobile antenna. Of these the most common are:

1. Capacitive shunt feeding. This is simply the addition of a shunt capacitor directly across the antenna feedpoint as shown in Fig 4.18. Capacitor values calculated by G3MPO are:

F (MHz)	29	24.9	21.2	18.1	14.25	10.13	7.05	3.65	1.9
C (pF)	18	27	37	74	150	300	544	1000	1000

Exact values can be determined experimentally and will need to be switched for multiband operation.

The way that this works can be seen by referring to Fig 4.19. The curve A represents the feed impedance of a Pro Am antenna in the frequency range 3.55 to 3.65MHz measured using the 3-m impedance box [7].

At the lower frequency the impedance is about R10 – 50jX, while at the higher frequency it is R70 + 70jX. On no part of the curve is the SWR better than 2:1. By increasing the inductance of the loading coil

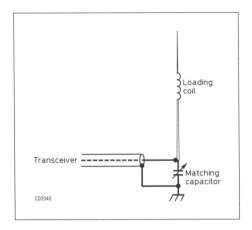

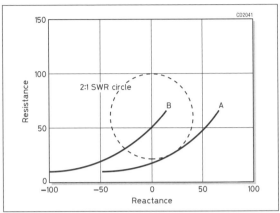

Above: Fig 4.18. Capacitor matching. In practice the variation in capacitance is achieved by switching in appropriate values of fixed capacitor

Right: Fig 4.19. Curve A – feed impedance of an 80m mobile antenna in the frequency range 3.55 to 3.65MHz. On no part of the curve is the SWR better than 2:1. An improved match is achieved by increasing the inductance of the loading coil slightly, and compensating with a capacitor across the feedpoint, thereby moving the curve to B

slightly and compensating with a capacitor across the feedpoint, the curve can be shifted to B to achieve an improved match.

2. Inductive shunt feeding. This is achieved with the addition of a small tapped inductance at the base of the antenna. With the loading coil adjusted to take into account of the effect of the base coil the antenna base impedance is raised in proportion to the size of the base inductance. There are two ways that this method of feeding can be implemented:

   (a) Selecting an inductor that results in a value greater than 50Ω at the lowest point of the antenna impedance/frequency curve. The transceiver connection is then tapped down the base inductance to obtain the best match as shown in Fig 4.20(a).

   (b) Using a variable inductance across the feedpoint as shown in Fig 4.20(b). The inductance value or tapping point must be changed when the frequency band is changed. A method of doing this is shown in Fig 4.21.

3. Transformer matching. This arrangement uses a conventional RF transformer wound on a toroid core as shown in Fig 4.22. A commercial or

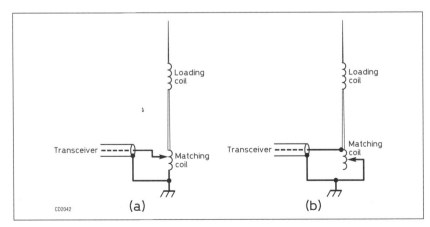

Fig 4.20. Two methods of using a tapped inductance for matching at the base of the antenna

CHAPTER 4: MOBILE ANTENNAS

homebrew matching transformer can be used. The one described is designed by G3TSO and uses toroid cores such as the Amidon T 157-2. It is wound with 20 bifilar turns using 18SWG (1.2mm) wire. Both windings are connected in series, in phase, and the second winding is tapped every other turn as shown in Fig 4.23. With the loading coil adjusted to take into account the inductance of the transformer windings, antenna impedances from 50Ω down to 12Ω can be matched.

The matching arrangement shown in Fig 4.23 is shown located partway between the transceiver and the antenna. In fact all of the matching arrangements shown could be connected in this way. As stated earlier, the feed impedance comprises the radiation resistance, earth resistance and coil resistance in series, which means that, in practice, the transceiver can be connected directly to the antenna without any matching at frequencies 14MHz and above.

On the lower frequency bands the coax feeder is so electrically short in a mobile installation that losses caused by a higher SWR are minimal.

If the matching arrangement can be adjusted from the driver's seat then this represents a greater degree of operator convenience than if it were located at the base of the antenna.

Fig 4.21. Practical implementation of the inductive shunt feed provided with the Texas Bugcatcher where a tapped inductor is connected across the feedpoint. The coil is fixed to the base of the antenna and the appropriate tap connected to the vehicle metalwork to provide the best match to the transceiver

VHF/UHF antennas

Manufacturers' catalogues and advertisements in amateur radio magazines offer a wide variety of VHF/UHF mobile antennas. Your choice will be affected by many factors, such as cost, method of mounting, performance and appearance. If you are interested in working DX, an antenna with some gain is helpful. If you are only interested in working through the local repeater then a quarter-wave antenna is usually adequate.

At 145MHz the quarter-wave whip's simplicity and limited height of about 480mm (19in) is often an accepted compromise. The physical dimensions at 70MHz (about 1.02m (40in)) and at 50MHz (1.48m (58.5in)) are such that size is the usual limit, making a quarter-wave whip preferable to a 'gain' antenna.

You can make a simple 2m antenna by

Fig 4.22. Matching arrangement for a mobile antenna using a variable ratio RF transformer

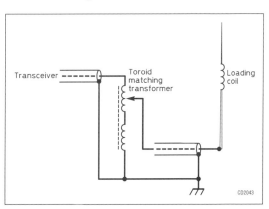

47

Fig 4.23. Details of the G3TSO RF transformer. (a) The circuit and impedance matching range. (b) Constructional details

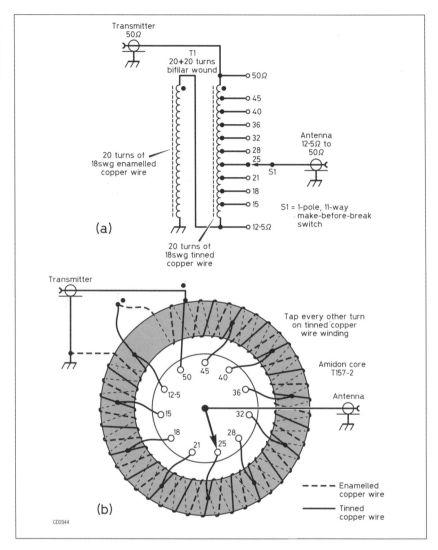

just soldering a 4mm plug to the end of 480mm length of hard-drawn copper wire; or even a wire from a metal coat hanger will do. Wrap some insulating tape (preferably coloured) around the other end to cover the sharp edge where the wire was cut.

If you are using a dual-band VHF radio then you will need an antenna for both bands. Should you use separate antennas for each band then you will need a *diplexer*, which combines the signals for both bands into a common feed line. If your radio does not have a built-in diplexer, using separate antennas not only saves you the expense of buying a diplexer but the combined cost of two antennas may be less than the cost of some dual-band antennas. However, you will have to mount two separate antennas.

On the other hand, if you prefer to mount only one antenna, you'll need a dual-band antenna.

An example of the VHF antennas I have collected over the years is shown in Fig 4.24. Starting from the right is a Mandol WT-1500 multiband antenna for 50, 144 and 430MHz bands. The antenna to the left is a lower section of ex-army whip antenna with a telescopic antenna soldered to it, which allows it to be tuned as a quarter-wave antenna for either the 50 or 70MHz bands. Ex-army whip antenna bases are very useful – they have a standard $3/8$in stud fixing so they can be used with any $3/8$in amateur antenna base.

The example shown in Fig 4.24 uses a PL-519 to $3/8$in adapter so that it can be fitted to my VHF antenna base. The other three antennas, from right to left, are all for 144MHz, being a quarter-wave, a $5/8$-wave and a vertical collinear respectively.

References

[1] *HF Antennas for all Locations*, 2nd edn, Les Moxon, G6XN, RSGB, 1982.
[2] 'A mobile antenna for 1.8–29MHz', M J Grierson, G3TSO, *Radio Communication* February 1988.
[3] 'An all-band antenna for mobile or home', John Robinson, G3MPO, *Radio Communication* December 1992 and January 1993.
[4] 'The "normal-mode" helical aerial', D A Tong, G8ENN, *Radio Communication* July 1974.
[5] *The ARRL Antenna Book*, 17th edn.
[6] 'Going mobile with a screwdriver', Peter Waters, G3OJV, *UK Radio Equipment Guide 2001*, Waters & Stanton.
[7] *The Antenna Experimenter's Guide*, 2nd edn, Peter Dodd, G3LDO, RSGB, 1996.

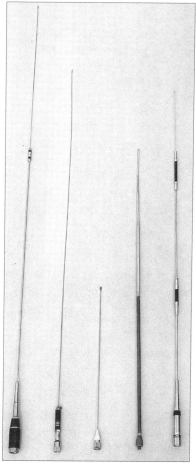

Fig 4.24. VHF antennas from the G3LDO collection – see text

CHAPTER 5

Fixing an antenna to a vehicle

Antenna location

One of the most frequently asked questions by anyone new to mobile operating is where to place the antenna on the vehicle. Conventional wisdom states that the best place to mount it is in the centre of the roof of the vehicle. While this is definitely the case for VHF I have never found any noticeable difference with different antenna mounting positions when it comes to HF.

In an attempt to resolve this question I have modelled the positions of an HF whip antenna on the front, top and rear of rather a crude model of a vehicle as shown in Fig 5.1.

For simplicity the vehicle is modelled as a small van with a 2.4m long antenna mounted on the top, lower rear and front. (In fact the rear-mounted antenna was increased in length so that the coil was clear of the roof of the vehicle in accordance with general practice). The roof-mounted antenna gave the best performance (as was expected) but not by a very great margin as regards gain, although it had a more symmetrical radiation pattern. As you can see there is a fairly high current in the metal structure of the vehicle where a high current carrying antenna element passes close by. It follows that when an antenna is radiated against the body of the vehicle, the vehicle itself becomes part of the antenna system.

The amount of RF current present in a vehicle structure was once demonstrated when a BBC camera crew were trying to film my mobile operation with the door of the vehicle open; RF attacked the audio system of the camera every time I transmitted.

The antenna current induced into the vehicle metalwork can be fairly high with some types of antennas and can interfere with the electronic engine management systems of certain vehicles.

To obtain a reasonably high radiation resistance a HF antenna should be made as long as practicable; this normally dictates that such antennas are mounted low on the vehicle.

Magnetic mount

This is better known as the *magmount*. It is a traditional method of fixing a VHF or a small HF antenna to the roof of a vehicle and was very popular

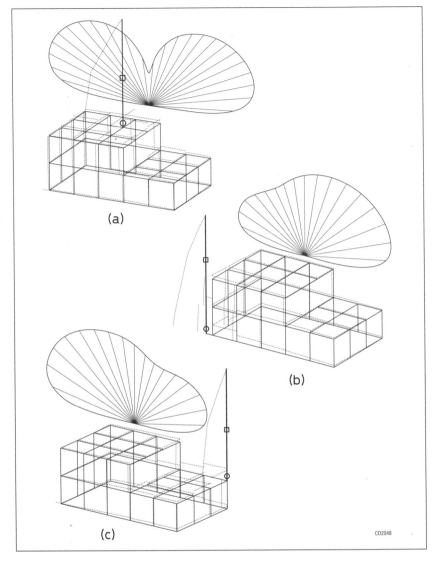

Fig 5.1. Computer model of an HF mobile installation with the antenna mounted on (a) top, (b) rear and (c) front of the vehicle. The circles and the squares on the antenna are feedpoints and the loading coils respectively. Also shown is the fore and aft elevation section of the antenna polar diagram. The vehicle is indicated by solid lines and the current with dotted lines. The distance between the vehicle line and its associated current line is an indication of relative current flow

with CBers. It uses an antenna base with a magnet to fix the antenna to a horizontal flat surface of ferrous metal vehicle body. It has the advantage of simplicity and requires no special fixing arrangements, which enables it to be used with a vehicle on a temporary basis if necessary. The coax feeder is routed either through the door seal (if you are using thin coax) or through a partly opened window (if you are using thicker coax). If you do route the cable through the window it is important that this is so arranged that opening a car door will not result in the magmount being dragged across the vehicle bodywork, damaging the connection to the magmount or the vehicle paintwork. Magmounts come with SO-239 sockets for VHF antennas or connectors for $^3/_8$in stud-mounted antennas for HF.

CHAPTER 5: FIXING AN ANTENNA TO A VEHICLE

Because magmounts are not directly connected to the body of the vehicle and rely on capacitive coupling, the antenna tuning may be different to other methods of mounting the antenna. At VHF this is not a problem because the capacitive reactance is low.

Magmounts come in many different sizes, as shown in Table 5.1, and are complete with antenna fixing and feeders.

For larger HF antennas the surface area contact between the vehicle body and magnetic base is insufficient to hold the antenna when travelling at any speed. You could use a much larger diameter magnetic base but this is rather impractical. The solution is to use three or four magnets held together in a frame. This arrangement spreads the load over a much wider area than a single magmount and can be used to support larger HF mobile antennas.

The most common multiple-magnet magmounts use three magnets. An example of one of these is the WMM-3401 series made by Watson (see Appendix 1) shown in Fig 5.2, which can support an average-size HF mobile antenna. It can also support a large antenna if operating from a stationary vehicle – see Fig 5.3.

Fitting and removing a magmount can be delicate operations. The trick is to use a lever to lower it in a controlled way when fitting it to the vehicle body. The magmount is removed by tilting it up and gently rolling it off in one smooth motion. Ensure that the surface of the magmount magnets and the vehicle body are free of dust and grit before placement to avoid any scratches to the vehicle paintwork.

Type	Diameter (mm)	Fixing	Feeder length (m)
Toyo KT-MG58	95	PL-259	4
Watson WMM-3.5	89	PL-259	4
Watson WMM-5	127	PL-259	4
Watson WMM-7	178	PL-259	4
Watson WMM-38	–	⅜in	4

Table 5.1. Example of available magmounts from W&S catalogue – see Appendix 1

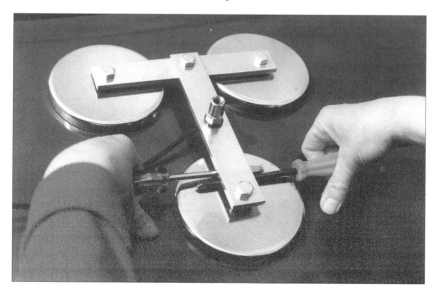

Fig 5.2. The WMM-3401 triple-magnet mount, comprising three 127mm diameter magnets held together in a frame. The magnets on WMM-3401 are very powerful and care is required when lowering the mount onto the surface of the vehicle. Here, the magmount is being lowered with the aid of two screwdrivers

Fig 5.3. A WMM-340 magmount being used to support a Texas Bugcatcher antenna. This antenna is too large for normal mobile when used with a magmount but is fine for static mobile operation

Through-panel mounting

Antennas can be fixed to mounting arrangements that require a hole drilled in the bodywork of the car – not the most popular method of fixing but it does have one great advantage and that is the fixing of the coaxial cable. The coax to the antenna can be run inside the vehicle right to the base of the antenna. Additionally the braid of the coax can be grounded right at the base.

If you mount the antenna on the boot lid make sure that it is placed so that the antenna does not collide with the roof when the boot is opened. Also use copper braiding to connect the boot lid to the body of the vehicle to improve the RF ground, especially at HF.

Most early commercial VHF radio installations used this type of antenna installation. The reason was that the transceivers were fairly large and the only suitable location was in the boot with a control cable to a small control unit that could be operated by the driver. This meant that the boot lid was the most obvious place for the antenna.

While I was serving as Force Communication Officer with the Sierra Leone Police Force (1964/65) I had the job of installing a VHF radio in the Governor General's new Rolls Royce. This proved to be a difficult task because of the robust construction and upholstery of the vehicle and the only way the supply and control cables could be run was by cleating them along the chassis, clear of the main bodywork. However, the antenna had to be mounted on the boot lid. The moment I stood with electric drill poised above the gleaming exterior of a brand new Rolls Royce is one I shall never forget.

The chassis bracket

On pre-1970 vehicles, antennas could be fixed to the rear bumper using chain-type or strap-type mounts. Modern vehicles very rarely have these convenient bumpers so a special mounting bracket is required.

CHAPTER 5: FIXING AN ANTENNA TO A VEHICLE

The chassis bracket is probably the best way of mounting a large HF antenna. It comprises a metal bracket fixed to the chassis or the rear floor area of the vehicle. If the vehicle has a tow bar with a bolt-on hitch, the simplest arrangement is to sandwich a steel plate between the hitch and its fixing, and fix the mount onto this adjacent to the hitch. An example is shown in Fig 5.4.

Ensure that an antenna mounted this way does not impede the opening of a tailgate or rear door – any arrangement that requires you to remove the antenna to open a door is clearly unsatisfactory.

I have made several of these brackets for the various vehicles that I have owned. The bracket is usually made from aluminium angle, the final design being determined by what is available at the local scrap yard – an example is shown in Chapter 6, Fig 6.5. Make sure that the corners of your bracket are rounded off and smooth – my first design caused the vehicle to fail the MOT test because of protruding 'sharp edges'.

Fig 5.4. An antenna mount fitted to the tow bar of G3ROO's car. The bracket comprises a section of 4mm steel plate sandwiched into the towbar. The antenna mount and ball-joint (to adjust the angle of the antenna) is installed on an angle mounting bracket, which in turn is fixed to the steel plate. The antenna is plugged into the quick-release fitting shown on the top of the ball-joint

The G3MPO antenna base

Most HF antennas use standard ⅜in connectors and most of the commercial fittings are designed to accommodate these. However, if you have made your own antenna then a home-made base fitting may be more appropriate. The following design by G3MPO used the chassis bracket described above.

The base comprises a short length of 22mm copper tubing insulated using plastic tubing and held in place with U-bolts, as shown in Fig 5.5. The lower section of the antenna is made from 15mm stainless steel tubing, which is held inside the 22mm tubing, the ends of which are modified to act as a quick-release mechanism.

The insulator is made from blue-coloured polythene tubing, which comes in a range of diameters, and which can be obtained from plumbing suppliers. A size is selected that is a tight fit over the 22mm tube. Two 38mm lengths of this plastic tubing are cut and fitted over the 22mm tube by either preheating the plastic tube or, if this is too difficult, by slitting it. The resulting gap in the insulator is electrically and mechanically unimportant. The insulator sections should be fitted so that there is approximately 25mm of copper protruding at each end.

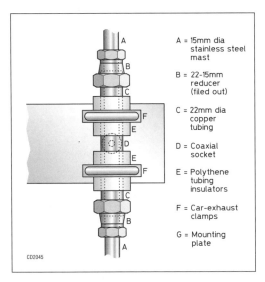

Fig 5.5. Details of the G3MPO HF antenna base construction

This insulated tube can be fixed to the mounting plate using exhaust-pipe U-clamps.

A 22mm to 15mm brass reducer is fitted to the top and bottom of the 22mm tube, which allows the antenna bottom section to pass right through the mount and be compression clamped into position. The shallow ring of brass machined into the bore of each reducer to prevent 15mm tubing going right through must be filed or drilled away. The brass sealing ring (called an *olive*) from the bottom reducer is not used and is discarded; the mast is clamped only by the olive in the top reducer.

On the prototype antenna a panel-mounting coaxial socket was fixed to the plate, and the centre pin soldered to the mid-point pigtail on the 22mm mount copper tube. A short coaxial cable goes from this connector through a small hole in the car bodywork, and then into the matching unit in the boot.

Luggage or roof rack antenna support

Many vehicles have special fitting arrangements for roof or luggage racks. A roof rack is a very suitable support for a mobile antenna – after all, if it is strong enough to carry luggage then it is strong enough to support a mobile antenna.

Most older vehicles had a built-in rain gutter around the roof and most roof racks at the time were designed to clamp on to this rain gutter. An example is shown in Fig 5.6, where the luggage rack is used to carry a variety of mobile antennas.

Modern luggage racks are made of metal, usually with a plastic covering and fixed to the vehicle with special fittings. A single bar of a luggage rack is a useful fixing point for a commercial antenna mount, as shown in Fig 5.7. To support a larger antenna the cross bars of a roof rack can be spaced so that they are a short distance apart and an aluminium plate fixed to them. This makes a good sturdy support for a larger antenna, as shown in Fig 5.8.

Because of the construction of the modern luggage rack it may be insulated from the vehicle body and some provision must be made for making a suitable RF earth.

One solution is to have a separate lead that runs from the base of the antenna to a suitable bare metal connection to the vehicle. Such connections are often difficult to find. One such connection point is one of the screws that hold the top safety belt fixture. If this lead is too long it will have some inductive reactance at the higher frequencies. Another way is to scrape away the paint by the door pillar, at a point where it is not visible when the door is closed. The base of the antenna can then be connected using a short lead, which is fixed to the vehicle with a self-tapping

CHAPTER 5: FIXING AN ANTENNA TO A VEHICLE

Fig 5.6. A photo of G3LDO mobile taken in 1980, near the home QTH! The bars of the rain gutter roof rack are placed about 300mm (12in) apart and an aluminium plate bolted to the bars. This plate is used to support antenna mounts for a helical HF whip and a 2m collinear antenna. Also fixed to this roof rack is a rectangular DDRR antenna for 14MHz – see Chapter 9 for more details

screw and a washer. Such a connection should be protected against corrosion with a thin film of grease.

Although you may think of ground connections having to be made with thick wire, in practice thin connecting wire is adequate and this can be routed via the door seals very easily.

Wire connections to the vehicle bodywork are fine for the lower-frequency bands but may have enough inductive reactance on the higher-frequency bands to complicate feeder-to-antenna matching.

One method is to use capacitive coupling to the vehicle. This comes automatically with a magmount but this may not be the ideal solution if you want to use a larger HF antenna. If you do have a luggage rack, then try a home-made magnetic coupler.

To make a magnetic coupler you need:

- Some magnetic material used to make removable signs on the side of vehicles
- A sheet of copper foil
- A short length of braid

One end of the braid is soldered to the copper foil then this foil is fixed to the magnetic material using vinyl electrical tape, as shown in Fig 5.9. The other end of the braid is connected to the antenna base, as shown in Fig 5.8. The coupler may be faced with thin plastic sheet if you are concerned that it may scratch the paintwork of the vehicle.

The size of the magnetic coupler will depend on the size and shape of the vehicle. It may be preferable to use two separate couplers, as shown

57

AMATEUR RADIO MOBILE HANDBOOK

Fig 5.7. Method of fixing an antenna mount for a VHF (or small HF) antenna to the single bar of a luggage rack

in Fig 5.8, rather than one large one for ease of handling. I recommend a minimum size of 300mm² (12in²)

The hatchback antenna mount

One of the most popular methods of mounting an amateur radio antenna to a vehicle these days is the hatchback mount. It has the advantage of being easy to fit and requires no modification to the vehicle. The only disadvantage is that it is the construction of such a mount limits it to VHF or a lightweight helix HF antenna. An example is shown in Fig 5.10 and availability is given in Appendix 1.

The bull-bar antenna support

Many four-wheel drive vehicles are fitted with a front bull-bar. This makes an excellent antenna support because of its location and strength and an antenna-mounting bracket is shown in Fig 5.11. The only drawback is that the antenna feeder has to be routed through the engine compartment. This is no problem with an old diesel vehicle (like mine) but there may be EMC implications for a modern vehicle; see Chapter 3. I have found this to be the most satisfactory of all places to locate an HF antenna. The main reason is that the antenna is always in view when driving, see

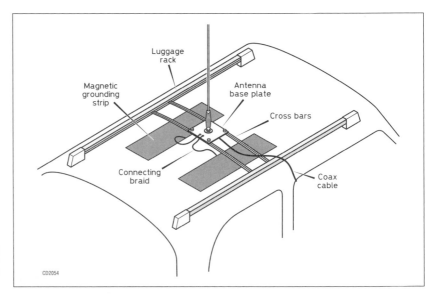

Fig 5.8. Luggage rack mobile antenna system with magnetic grounding pads

CHAPTER 5: FIXING AN ANTENNA TO A VEHICLE

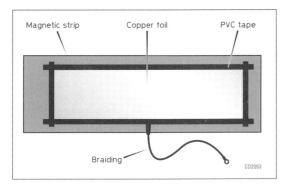

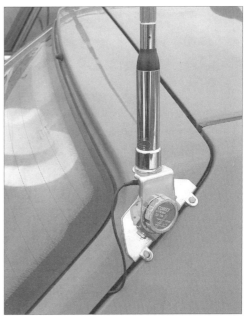

Fig 5.12, and it is easier to avoid any altercation with low tree branches and vehicle park height restriction devices. I once drove out of a car park after an operating session and wrecked my HF antenna on a height barrier because I had forgotten to take it off. If the antenna is on the front of the vehicle this type of event is less likely to happen.

Coaxial cable antenna feeder

All mobile antennas require a coaxial cable feed. I have always used RG58BU because I use it for all coax interconnections around the shack. Furthermore, being only 5mm in diameter, it is easy to route through door seals and other obstructions encountered in mobile installations.

With my two previous vehicles I have run the coax cable from the radio to roof-mounted antennas via the inside of the windscreen pillar; the coax emerged to the outside world via a hole at the top of the door pillar. An example of this arrangement can be seen in the feed details of the roof rack antenna shown in Chapter 9.

Routing a length of coax this way is achieved by first pushing a length of hard-drawn copper wire through the hole in the door pillar and feeling its way down the windscreen pillar. The coax is then fixed firmly to the copper wire and gently pulled back up the window pillar. This procedure should only be attempted if you have a lot of patience.

Some of you may question the losses of such a choice of feeder, especially at VHF/UHF.

The loss quoted for RG58 is about 1.8dB for 30m at 30MHz. The loss for a 3m length, such as may be found in a mobile installation, is around 0.18dB at 30MHz and just over 1dB at 400MHz.

G3YXM uses very thin RG174, which is even easier to route round

Left: Fig 5.9. A magnetic ground coupler
Above: Fig 5.10. Hatchback antenna mount being used to support a 50MHz antenna on the rear of G0GCQ's car
Below: Fig 5.11. Home-made bullbar bracket using a small piece of aluminium angle. The bracket is connected to the chassis of the vehicle using thick copper braid with a plastic protective sheath

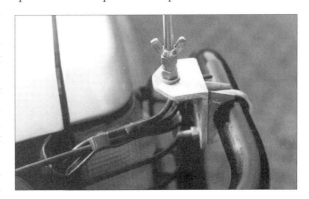

AMATEUR RADIO MOBILE HANDBOOK

Above: Fig 5.12. Texas Bugcatcher mounted on the bull-bar of a four-wheel drive vehicle

Right: Fig 5.13. The Kingdon cable kit, which provides a low-loss feed to the antenna. A short length of RG174-type cable is used so that it can be routed via a door or hatch seal

obstructions and through crevices, having an outside diameter of only 2.8mm. For a 3m length the loss will be around 0.7dB at 30MHz. Power handling is not a problem with a matched antenna; G3YXM runs 400W at HF.

I have a coax cable kit on my vehicle, which has the tradename 'Kingdon' marked on the side. It comprises 5m of 5D-FB low-loss cable and about 400mm of what looks like RG174. The RG174 end is terminated with an SO-239 connector and the other end of the cable is fitted with a PL-259 plug. A VHF/UHF installation using this kit is shown in Fig 5.13.

A very similar cable kit, labelled 'W-ECH', is made by Watson; see Appendix 1 for availability.

CHAPTER 6

Kite and balloon antenna supports

You can get some spectacular results with mobile or portable operation on the low-frequency bands if you use a kite or a balloon. If you live in an area where there is hostility to large low-band antennas, operation from a vehicle at a remote site is an option.

The antenna can be a long wire fixed to a tall tree or supported by a kite or balloon. You can use your existing HF mobile equipment without any change and build a simple ATU that fits to the existing antenna mount. This chapter describes some of the considerations for operating DX on the 80m band, some of which could also apply to 160m. Most DX openings on these bands are early in the morning or after dark. Fixing antennas to trees in the dark is not my idea of fun so this chapter concentrates on using kites and balloons as a support for a wire antenna.

Air traffic control regulations state that the maximum height a kite or balloon can be flown without notification is 60m. Because of the angle at which the kite flies relative to the anchor point you can use an antenna longer than 60m. In practice I find a 40m (130ft) antenna is adequate, which gives a half-wave antenna on 80m and a quarter-wave one on 160m.

Kites or balloons should not be flown anywhere near overhead power lines or over buildings or roads. Never fly a kite during electrical storms or if the weather is severe in any way.

Kites

A kite is a good way of supporting a long-wire antenna, provided the wind strength is in the range of light breeze (4 to 6 knots) to fresh breeze (17 to 21 knots).

The essential requirements for a kite used for this purpose is that it is simple, easy to launch, provides a good lift at low wind speeds and is rugged. When it is flown it should be stable – by this I mean that it should remain stationary in the sky even in turbulent wind.

A kite with these characteristics is the *delta* shown in Fig 6.1. The construction is shown in Fig 6.2. It is constructed from a tough lightweight material called *ripstop nylon* and uses fibreglass spreaders. The side and centre spreaders are fixed into the kite when it is constructed. The cross

AMATEUR RADIO MOBILE HANDBOOK

Fig 6.1. A delta kit supporting an antenna. The ribbons at the wingtips are stabilisation streamers

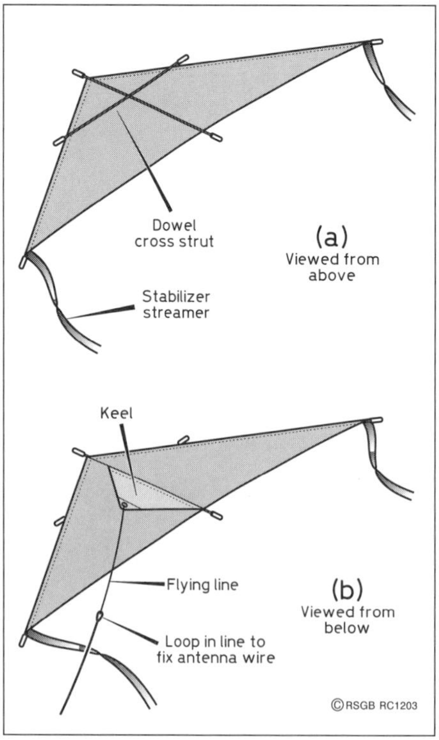

Fig 6.2. Construction of a delta kite

spreader is fixed to the kite when it is prepared for flying and removed so that the kite can be folded for transportation. When the kite is flying the cross spreader flexes, increasing the dihedral, as can be seen in Fig 6.1 – this reduces lift and increases stability in strong gusts of wind. Details of how to construct such a kite is given in reference [1] although suitable kites can also be purchased (see Appendix 1).

I find that an essential item of equipment for launching and recovering kites is a winder, as shown in Fig 6.3. This allows the antenna and the line to be played out and recovered in a tidy fashion, a major consideration when launching a kite in darkness or half-light.

The antenna I use comprises 0.6mm plastic covered wire wound around the thin nylon kite flying line. The method of connecting the antenna to the ATU will depend on the layout of the equipment, existing antenna system and suitable tie points. A couple of methods are described below.

If you have a vehicle with a roof rack then this provides an excellent anchor point. The flying line is fixed to the vehicle as shown in Fig 6.4. A length of bungee cord can be used to reduce the strain on the line and kite in strong gusty winds. The wire end of the antenna is then brought through a wind-down rear window using plastic tubing to form additional insulation.

If you do not have a roof rack a door pillar may be used as the anchor point.

When the wire is in position the window can be wound up, making an effective insulated clamp. The antenna wire must be arranged so that it is always slack – whatever the position of the kite, the nylon flying line must take all the strain.

A different method of connecting a kite antenna to the antenna radio is shown in Fig 6.5. This uses the existing antenna mount used for normal mobile operation. The ATU is fixed to the mount and the low-impedance feed from the original antenna socket is then connected to the ATU. The antenna wire is then connected to the high impedance point on the ATU, as shown in Fig 6.6.

Balloons

The wind often disappears in the early dawn or after sunset, when the DX is at its best, just when you needed a breeze to keep the kite antenna flying. A solution to the problem is to use a balloon. Mylar balloons are often obtainable from flower shops. The largest I was able to obtain was 1m in

CHAPTER 6: KITE AND BALLOON ANTENNA SUPPORTS

diameter and was referred to as a *jumbo balloon*. They have to be transported in an inflated condition if you want to conserve expensive balloon gas.

At least three of these balloons (preferably four) are required to lift a half-wave 80m wire. An anchor point to the balloon is made using brown plastic tape as shown in Fig 6.7. Balloons can be harnessed together via the plastic tape loops with a nylon cord and the antenna wire tied to the harness.

The pull is very much less than a kite in no-wind conditions when a balloon can be flown. The main disadvantage of using a number of relatively small balloons linked together is that the lifting system has a lot of drag. The effect of drag is that even a very light breeze will blow the balloons sideways so that the antenna is anything but vertical. This defeats the object of using a balloon to make a large vertical DX antenna.

Ideally, a balloon should be streamlined, like a barrage or advertising balloon, to increase the lift-to-drag ratio. Nevertheless, a cluster of balloons is useful when the wind is zero on the Beaufort wind scale (wind speed of 1 to 3 knots). The antenna must be as light as possible so a thin wire (0.6mm plastic covered), without the nylon flying line, can be used. Fine aluminium welding wire in parallel with fishing line can been used as a possible alternative.

Balloon gas is about 98% helium and 2% air. This gas comes in hired 65cm tall cylinders, which are not heavy and are easy to move. The cylinder and gas can be obtained from regional depots – see the *Yellow Pages* for information on a gas supplier near you. A cylinder will fill about 12 balloons although in practice you only need to fill two and top them up weekly with a small amount of gas.

Choosing a site

A good location for flying a kite is dictated by interaction of the

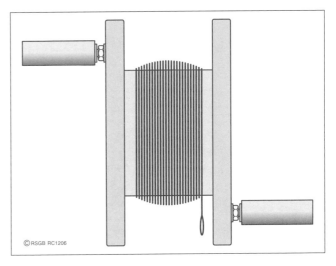

Fig 6.3. Flat wooden winder for launching and recovering kites or balloons

Fig 6.4. Method of connecting a kite or balloon antenna to a vehicle with a roof rack. The bungee cord is not required for balloons and is only of use with kites flown in blustery winds

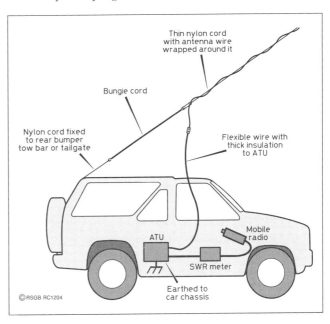

Top: Fig 6.5. A home-made ATU fixed to an existing mobile antenna mount. The mount itself is made from three sections of angle aluminium, which form a complicated structure to circumvent the plastic bumper. The ATU is designed specifically to feed a high-impedance 80m antenna

Middle: Fig 6.6. Circuit diagram of an ATU for feeding a high-impedance (end-fed half-wave) antenna. The values of the inductor and capacitor are not critical. The prototype was built on a flat piece of aluminium

Bottom: Fig 6.7. Method of connecting an antenna to a balloon. The nylon cord is used for tethering two or more balloons together and the antenna is connected to the cord

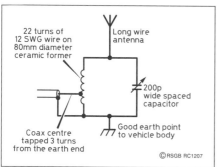

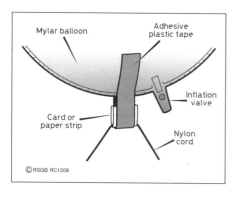

ground with the wind. The best location is a flat open space with no buildings or trees, so that regardless of which way the wind blows the wind turbulence is minimum. Flat seaside shores are good, particularly with an onshore breeze. This type of location should also theoretically provide a good radio QTH.

Hilltops, while being fine for VHF operation, are not ideal for flying a kite. This is because there is a lot of turbulence and down drafts on the down-wind side of the hill.

I normally operate from Southdown, near Worthing in Sussex, because it is conveniently close to home. It has a public car park about one-third up the Southdown hills facing the sea. If there is a southerly on-shore breeze then I'm in luck as air moving towards a gentle slope gives good lift – to a kite or balloon. If the wind is blowing from the north or north-east then operation is difficult or impossible.

Keep kites and balloons away from obstructions such as buildings and trees. These obstructions create a lot of air turbulence as well as providing a trap for the kite or balloon.

I use a separate antenna and winder for the balloon antenna system. If the wind speed increases I then wind in the balloon antenna and launch the kite antenna.

Reference

[1] *Kite Cookery*, Squadron Leader Don Dunford, MBE. See Appendix 1.

CHAPTER 7

Bicycle HF mobile

Fig 7.1. The G3LDO/M HF bike mobile. The rig is located in a handlebar-mounted bag, which also gives good weather protection to the radio and provides stowage for the key, headset and logbook. A thick aluminium plate is fixed to the rear carrier to provide a place to fit the battery and a mounting point for the antenna

The easiest way to go bicycle mobile is to use a VHF/UHF handheld. However, I find HF operating from a bicycle presents the greatest challenge and gives the most satisfaction. HF mobile from a bicycle works surprisingly well, with SSB contacts over a distance of 2000km presenting no difficulty at all on the higher HF bands. But be prepared for some strange glances from other road users!

My bikemobile comprises a standard mountain bike with front suspension forks; the latter reduces modulation of the voice when travelling over rough ground – see Fig 7.1. A further modification is a rear carrier on which is fixed a thick aluminium plate – this provides a place to fit the battery and a mounting point for the antenna. I use any one of the helical-wound antennas described in Chapter 4. In general I use the Q-Tek because it is more compact and multiband. The rig is simply placed in a handlebar-mounted bag, which was obtained from a local bike shop – no special fitting needed at all. This arrangement gives good weather protection to the radio and provides stowage for the key, headset and logbook.

There are a lot of small rigs available on the market now for this sort of operation. I use an IC-706 Mk1 because I happen to have one. The power is turned down to position 2, which gives approximately 10 to 15W. With a 12Ah lead-acid battery this gives me just over one hour of transmit operating. If conditions are very good then you can reduce the power down to 'L' (about 5W) to extend operating endurance.

The IC-706 has a reputation for low audio. This was fixed using a Heil HS706 headset with a boom microphone. This headset has an internal microphone amplifier and is made especially for the IC-706. When

Fig 7.2. Keying is done with a simple lightweight paddle fixed to a wood base, which is wedged in the carrier with the rig. This photo was taken just after a QSO with PR7NJ

operating on SSB using the VOX contacts around Europe and occasional DX contacts can be made very easily while riding the bike. I make most of my contacts cycling along the beach when the tide is out. The saddle is set low so that I can stop with both feet comfortably on the ground so that the logbook can be updated or any appropriate notes made.

I also use CW when operating from the bikemobile, although I am not clever enough to do this while on the move. Keying is done with a simple lightweight paddle fixed to a wood base, which is wedged in the carrier with the rig as shown in Fig 7.2. For DX operating in less-than-perfect propagation conditions this is definitely the preferred mode.

Due to the limited endurance of the bikemobile set-up, I limit my DX operating to a combination of reasonable weather and conditions – and the tide has to be out far enough to expose enough sand for riding the bike. The battery is recharged from a charger inside the garden shed.

Doubts have been cast as to the quality of a bicycle as an antenna counterpoise. I have made several antenna measurements; it seems that any antenna (in the frequency range 14 to 30MHz) resonated on a large vehicle maintains the same resonant point when the antenna is fixed to the bicycle. Furthermore antenna resonance is not changed when I am riding or just standing by the bike sending CW. With antennas below 14MHz it is a different matter. The resonance changes if I touch the bike frame when transmitting, which indicates that the bike is not as an effective ground plane as at the higher frequencies.

I use my bike on a daily basis as a mode of transport. As a result it is fitted with items that are not normally seen on mountain bikes and are not regarded as fashionable accessories. For example it has mudguards so I don't get covered in muddy water when it rains or salty sand when I cycle on the beach. It also has a rear view mirror so that I know what is going on around me and avoids the inevitable wobble if you have to look round to turn across traffic. Additionally, it has a bell – some of the paths around here are shared by bikes and pedestrians.

Because I don't go racing down rough twisting mountain tracks at a high speed (what the mountain bike was originally designed for) I don't wear a cycle helmet. As a normal piece of headgear it doesn't keep out the rain off when it is wet or the sun when it is hot – so I go for something more practical, as do most of the other commuter cyclists around here.

However, having said all that, I do recommend that you wear a helmet if you cycle a lot in heavy traffic or you cycle cross-country on rough terrain.

Fig 7.3. The most unusual feature of the VE3JC installation is the aluminium tube extending from the bike frame and rear pannier rack, which allows the rider to mount and dismount in the conventional way without conflict with the antenna. The rig is placed in a handlebar bag

The VE3JC bikemobile

John Cumming, VE3JC, is a dedicated bikemobiler. He combines QRP with one of his other favourite hobbies – cycling. The complete bikemobile set-up is as follows:

- One very-well-used 18-speed bike (a Canadian Tire special purchased about 12 years ago for under $200).
- Index Labs QRP+, original version with no internal modifications, usually running 4 to 5W on CW and 2W on SSB.
- Outbacker Perth 75 to 10m antenna
- 7Ah sealed lead-acid battery.

The complete station is shown in Fig 7.3. VE3JC uses the same method of mounting the rig as I do – just putting it into a handlebar bag, angled so the front of the rig is visible and the controls accessible while cycling. A closer view is shown in Fig 7.4. He does not use earphones; instead a 'deflector' is used to direct the audio from the top-mounted speaker to the rider (like 'His Master's Voice'). This arrangement allows surprisingly easy copy even in the presence of cars and cows.

The antenna is mounted on an aluminium tube extending from the

Fig 7.4. The QRP+ 'installed' in the saddlebag and the miniature paddle key attached to the right extension of the handlebars on the VE3JC bikemobile

bike frame and rear pannier rack to about 300mm (12in) behind the back wheel. This arrangement allows mounting and dismounting in the normal way without touching the antenna. This overcomes the problem I have with my bike where I get tangled up with the antenna unless mounting and dismounting is planned with care.

The bend at the end of the extension tube permits mounting of a super-bright-LED flashing light. This light unclips and makes a great night light for portable operation from a tent.

The aluminium tube attaches to the bike frame using two plastic U-lock mounting brackets (50 cents each at a local junk store). The aluminium mounting pole is anchored to the rear pannier rack using an 'inverted' mobile antenna bracket and a good old Canadian hockey puck. Removal of the whole antenna system from the bike takes only 30 seconds: undo the butterfly nut on this anchor and slide the pole assembly back (U-lock brackets on bike frame remain permanently attached). The original aluminium pole was cut from a folded dipole element of an old commercial antenna; a more readily available source would be a UHF TV antenna support mast; these come in all sorts of shapes so one can be selected that is suitable for an extension.

The antenna stinger [1] must be extended to almost maximum to achieve resonance on most bands. Although there is some sway in the antenna this is hardly noticeable.

Changing bands is no problem – a 15-second stop to change taps on the Outbacker, and back on the road again.

The mini-paddles are mounted to the right extension of the handlebars. Some people have used contact switches in each handlebar extension (dit with left thumb, dah with right thumb) which may be marginally safer but has not been tried. This allows the wrist to remain on the handlebars when transmitting, so steering, shifting and braking are still convenient; however, it is difficult to send good CW when riding on a rough gravel road.

The dial on the QRP+ is surprisingly stable. The frequency does not change even with severe bouncing on bad roads in spite of having no front suspension on the bike. Bikemobile QRP is not for those who like to keep their rigs in mint condition.

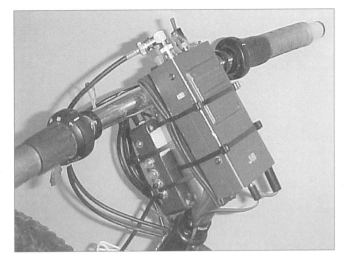

The KB8U bikemobile

Russell, KB8U, has an interesting bikemobile set-up. It comprises a Mizuho MX-14S SSB/CW 20m handy-talky fixed to the handlebar stem with cable ties with four layers of rubber from an old inner tube for shock absorption as shown in Fig 7.5. A homebrew electronic memory keyer is fixed to the bottom of the stem.

Fig 7.5. The Mizuho MX-14S SSB/CW 20m handy-talky and homebrew electronic memory keyer fixed to the handlebar stem on the KB8U bikemobile

The rack has a flat piece of metal fixed to it with a hole for the antenna mount. The antenna is a commercially centre-loaded fibreglass whip (just over 2m long). The rig is powered using 11 4400mAh NiCd batteries in an aluminium box strapped to a rear rack – see Fig 7.6. It has been modified with a 15W amplifier installed in the rear battery compartment. Cable ties are used to attach the rig, wires and battery box to the bike frame.

Russell is another of those clever guys who can send and receive CW while riding a bicycle.

He has operated HF CW bicycle mobile occasionally for several years, but never found a comfortable arrangement that allows keying while at the same time having a good grip on the handlebars. However, one arrangement that works is shown in Fig 7.7.

Fig 7.6. The box fixed to the carrier on the rear of the bike contains the batteries and a 15W PA. This is fixed to a plate, which in turn is fixed to the carrier to support the box and the antenna-mounting unit

The key is made from a small piece of double-sided circuit board material measuring about 4 by 1.5cm (1.5 by 0.5in). A cut is made in the copper cladding in both sides of the board. A piece of brass shim stock is then soldered to each side of the board to make up the dot and dash switches. The shim stock is bent very

Fig 7.7. The home-made paddle key and key mounting used by KB8U

slightly out so that it doesn't short-circuit unless the shim stock is depressed against the circuit board.

The key is mounted by modifying the Grip Shift SRT400-32 front changer; a groove is cut in it into which the circuit board is then jammed. The key is positioned such that it is not accidentally knocked while shifting gears but is easy to use while gripping onto the shifter and handlebars. The key was made as narrow as possible to minimise the difference between the hand position used while sending code and normal grip on the handlebar. The shifter is rotated on the handlebars so that the key is closest to where the thumb and index finger are if those fingers are not gripping the bar.

KB8U is able to send fairly good code even on moderately bumpy gravel roads and yet have a safe and comfortable grip on the handlebars with his hand close to the brake lever. Only the thumb and index finger are in a different position than normal, and they do not tire as quickly as other spots tried to mount a key.

The disadvantage of this installation is that it is not at all waterproof, so some provision is necessary to cover up the rig in bad weather.

Note

[1] 'Stinger' is the USA term for the part of the antenna above the resonating coil, the length of which can be adjusted to resonate the antenna.

CHAPTER 8

Maritime mobile operation

Maritime operation is very popular with licensed operators who own boats. In addition to the normal amateur activity, such as the pursuit of awards, DX chasing and QRP operation, there are maritime nets in many countries around the world on 14.300MHz or 14.320MHz.

At these meeting places you will hear enquiries about the weather, position reports, marina availability and charges, and other topics of mutual interest.

Additionally, you will often hear maritime operation from a ship or container vessel where a member of the crew has an amateur radio station.

Installing HF radio equipment

Fitting HF radio to a small boat is similar in some respects as fitting HF radio to a vehicle, particularly on a powerboat where there is no rigging. However, on a boat there is generally more room to install radio

Fig 8.1. Laurie Mayhead, G3AQC, operating from his boat 'Innisfree'. The TS-430 is mounted, together with a FAX-1 weather fax system to receive weather charts on 512kHz, in a stowage compartment above a bunk. This arrangement keeps the equipment dry whatever the sea conditions

AMATEUR RADIO MOBILE HANDBOOK

Fig 8.2. A commercial installation of an IC-M270 MF/HF transceiver on board 'Muhuhu'. The main transceiver is shown located in the radio/navigation instrument space with associated electrical distribution wiring

equipment, although the installation will have to be done to a high engineering standard to ensure that it is weather and waterproof. Where the radio is installed depends on the layout of the interior of the boat and the required operating convenience.

The transceiver should be sited where it can be operated conveniently, yet at the same time be protected from splashes of salt water. Fig 8.1 shows the installation of an amateur transceiver in a stowage compartment.

The other two important considerations are a power supply and an antenna. The type of radio installation will depend a lot on provisions already in place for radio equipment and whether the rigging has already been modified for use as an antenna. If you are contemplating a temporary installation then a temporary antenna can be used, but it must be rigged in such a manner that it does not interfere with the working of the sails. These considerations are described in more detail later.

Power supply connections

Most amateur radio transceivers run on 12V and, as most boat power supplies are also 12V, obtaining power to run the transceiver normally does not present a problem.

Bear in mind that the power supply cable diameter must be large enough to carry the 20A peak current without any appreciable voltage drop – cable sizes of 6 to 10 mm^2 are fairly typical. The method of connecting to the battery is similar to that described in Chapter 3.

Some transceivers, such as the IC-706, can be installed with their front panel remote from the main transceiver. In this case the transceiver can be installed in any available space as shown in Fig 8.2 so that it does not take up valuable space at the chart table.

Antennas
Backstay antenna
The most popular and effective way of making an all-band HF antenna on a sailboat is to make use of a piece of the boat's existing steel wire

rigging. Although the stainless steel rigging wire does not have the conductivity of copper, the diameter and length of the stays make them very usable radiators.

The most practical stay to use is the one that runs from the head of the mast to the stern of the vessel, known as the *backstay*. Even on a moderately sized family sailboat, this stay is over 10m (30ft) long. The most common method is to have insulators inserted into the metal backstay to insulate it from the mast and the rest of the rigging so that it can be used as an antenna. If your boat has not been modified in this way and you are considering a permanent radio installation then fitting these insulators would be the way to go. However, it is important that the job of fitting these insulators is done to the highest standard because the strength (or weakness) of the backstay affects the integrity of the sail rigging as a whole. Unless you know what you are doing, modifications to the backstay should only be carried out by a competent boatyard contractor.

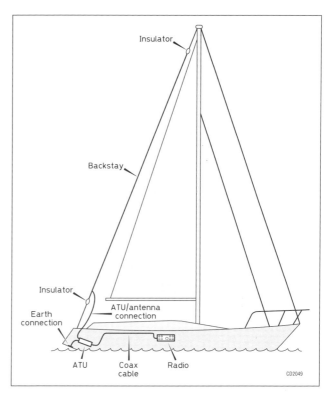

Fig 8.3. An overall view of an HF radio installation using a backstay antenna

An overall view of an installation using a backstay antenna is shown in Fig 8.3. The uppermost backstay insulator should be at least 1m (3ft) from the mast attachment point. The lower one should be sufficiently above the deck to ensure that the antenna portion of the stay is clear of most of the seawater that floods the deck in heavy weather. Additionally it prevents accidental contact when it is being used as a transmitting antenna.

These insulators are expensive. A less elegant (and less expensive) solution is to insert, in the same locations as described above, ceramic egg strain insulators as shown in Fig 8.4. The method of fitting these insulators to the stay are shown in Fig 8.5.

The exact length between the insulators is not critical, since the antenna will need an ATU and be operated as a vertical random wire. As a rule of thumb it should be less than half a wavelength at the highest frequency you intend to use. In practice this would be around 9m. This length also avoids multiples of half-wavelengths on any of the frequencies that you intend to operate on, thereby avoiding the wildest extremes in impedance. The advantage of the random wire approach is that it can be made to operate on all amateur and marine frequencies with an ATU. This type of antenna requires a good counterpoise, which is described later.

Fig 8.4. An insulator designed specifically for insulating sections of stays. The other two insulators (known as *egg insulators*) are similar to those used to insulate the guy support wires of small power line poles

Fig 8.5. Method of connecting insulators to a stay. (a) Egg insulator and (b) commercial backstay insulator

Although the backstay is normally used as an antenna, one of the sidestays can be used. The advantage of this arrangement is that it may come very close to the where the radio is installed, in which case a manual ATU, situated close to the operating position, can be used.

Antenna tuning unit (ATU)

An essential part of using the rigging as an antenna is an ATU (or tuner), which should be located as close to the antenna as possible. Because this arrangement often requires that the ATU be located in the stern, frequently in an inaccessible space during normal use, an automatic tuner is normally used for this purpose. The suitable automatic tuners are contained in a semi-waterproof housing that will allow them to be mounted in a fairly sheltered location right at the base of the antenna. They are connected to the radio by standard coax and usually a 12V DC cable. An example of this type of installation is shown in Fig 8.6.

The wire running from the ATU to the antenna side of the insulator should be at least 8 gauge, with high-voltage insulation suitable for exterior use. The wire used to supply the high voltage to neon advertising signs can be used, although coax cable with the outer sheath and braid removed is more readily available. If modified coaxial cable is contemplated for a connection between the ATU and the antenna, use RG213, which has an adequate insulation thickness and has a tough multi-wire centre conductor. Additional insulation can be added by enclosing the antenna wire in plastic tubing. A correct-sized waterproof gland must be used where the cable passes through the deck, as shown in Fig 8.7.

If a sidestay or a temporary wire is located close to the radio installation then a manual ATU can be used, as shown in Fig 8.8.

Counterpoise

Copper foil is used on most non-metallic boats to provide a good RF ground, or counterpoise, which is necessary when using an end-fed wire such as a backstay or sidestay. A continuous run of 75 to 100mm wide, 5mm or better, foil should be run from the ATU to the engine earth system and keel bolts where appropriate.

If the boat's anti-electrolysis bonding measures are correct, the engine block should be connected to all metal through hull fittings. With this type of installation the copper grounding foils should be attached to the

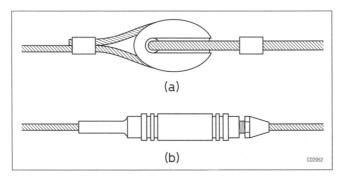

CHAPTER 8: MARITIME MOBILE OPERATION

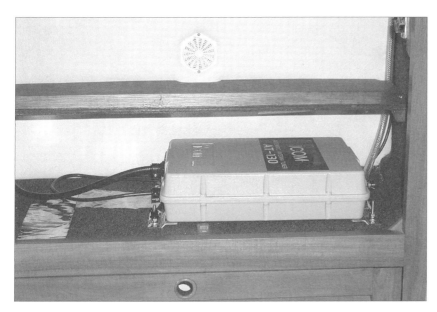

Fig 8.6. An ATU fitted in a stern stowage compartment showing the antenna wire, with additional PVC tubing insulation to the right and a copper strip to the left, which is part of the counterpoise system

lug on the tuner as shown in Fig 8.9. A second length of foil should be run from the operating position to the same point and connected to the radio's grounding lug.

The foil should be laid flat to the hull and can be secured with contact cement. Corners can be turned by folding the foil rather like a mitred corner. Try to run a continuous length. After installation, it can be painted over if required.

Fig 8.7. Shows the insulated antenna connection from the ATU below deck to the backstay on G3AQC's 'Innisfree'. The antenna wire is fed through the deck using a waterproof gland. The backstay on most wood or fibreglass boats is not connected to the ground system so the wire can be taped against the backstay

AMATEUR RADIO MOBILE HANDBOOK

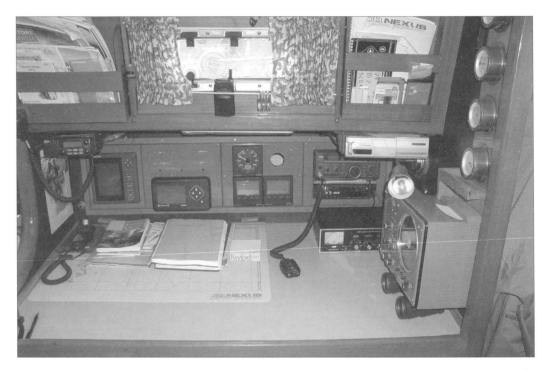

Fig 8.8. The chart table on board 'Hellene Louise' owned by Chris Langmaid. The HF amateur radio installation is shown in the right-hand corner of the chart table with a temporary manual ATU. In the foreground (right) is an older HF marine radio

The centre-fed antenna

A centre-fed antenna is also a useful antenna, particularly if no provision has been made for the rigging to be used as an antenna. This antenna is often thought of a single-band device but, used with an ATU, a 7 or 8m centre-fed length of wire should work on all bands from 7 to 21MHz. Of course you could use a half-wave length of wire (dipole) without an ATU as a single-band antenna and, as mentioned in Chapter 1, most /MM activity takes place on the 14MHz band.

The connection from the ATU or radio to the antenna feedpoint should be made with a good grade of RG213 or better coaxial cable. In the multiband configuration, there will at times be quite a high standing wave ratio on the feeder but, at the usually short runs involved, the losses will not be significant.

The feeder connection to the antenna must be carefully sealed because any salt water that is allowed to wick along the braid will render it useless – see 'Antenna connections' below.

HF whip antenna

The random wire antenna with an ATU can also be used on boats where the rigging is unsuitable or non-existent, such as powerboats. Here a marine-grade Fiberglas IM whip can be matched with the tuner and the same grounding requirements are used as described above. These whips are in the range of 7m to 11m long, require sturdy mounting flanges, and can put a significant dent in your wallet.

There are two ways in which a whip antenna can be used. One is to have a whip that is resonant at the highest frequency that will be encountered and to use a remote ATU to resonate it on the lower frequencies. The other is to use a single-band resonant whip without an ATU.

Whip antennas used for marine use should be inspected carefully if they are to be installed permanently aboard a boat. Antennas made from chromium-plated steel or zinc/aluminium alloys should not be used.

Antenna connections

The attachment of both the single-wire feed and the coax to the antenna feedpoint must have a very good electrical and mechanical connection. In addition measures must be taken to prevent ingress of water into the cable. Electrolytic action of dissimilar metals must also be minimised and the vast majority of installations make use of either bronze split bolt and nut combinations, available from electrical suppliers, or small all-stainless-steel hose clamps. Be sure to check the bolt on the supposedly all-stainless clamps sold in discount hardware stores as these are often *not* stainless.

To attach the wire to the stay use two, or preferably three, clamps or nuts spaced around 50mm apart on each leg of the stripped coax, or just three on the single wire. The e-clamps and joint area should be carefully and liberally coated with marine-grade silicone sealant, taking care not to leave any pinholes or gaps where water could he trapped and cause a noisy connection.

Fig 8.9. The ATU fitted into the stern section of the hull, aft of the steering mechanism on 'Innisfree'. The earth lug of the ATU is connected to the grounding foil although the foil is difficult to see because it has been painted over

Seagoing RF

On a merchant ship the antenna is usually well elevated and the operating position is adequately screened, but on a yacht the reverse is true and everything can easily become 'humming' with RF. Even if a linear amplifier can be powered, the stray resonant circuits provided by the stays, rigging and wiring on a yacht can be positively dangerous when excited by a strong RF field. It is therefore far better to keep the power

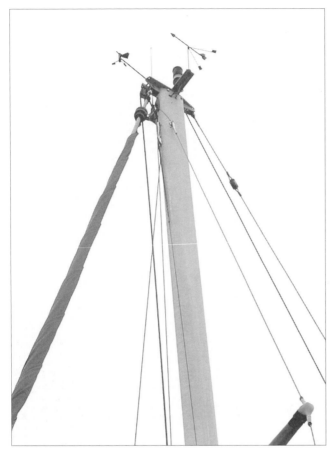

Fig 8.10. The top section of the mast on 'Innisfree'. The VHF antenna and weather instrument sensors are placed on brackets to that they limit the angle of obstruction to the navigation light. Note also the top insulator on the backstay, which is used as the HF antenna

level down and make a really serious effort to get maximum radiation and minimum reflected power. Even with low power, other electronic equipment on board may seriously misbehave when subjected to RF. Electronic speed/distance logs and quartz crystal clocks should be carefully checked out for RF tolerance before setting out on any major voyage.

VHF

Most boats have marine VHF radio fitted as standard. The marine band is a group of internationally standardised frequencies between 156 and 162MHz, which is pretty close to the amateur 2m allocation. The VHF marine installation also has a half-wave or vertical collinear gain antenna and this may function very well on 2m. The amateur 2m band is only 8% off the resonance of the marine antenna and the SWR might be a bit high. However, your radio should handle the small mismatch and it is worth trying this possibility before going to the trouble of installing a separate amateur radio antenna. You will need an antenna switch to change the antenna from the marine radio to the amateur rig if you use this approach. You will need a separate antenna for any other amateur VHF band.

A different solution to the antenna problem might be to install a multiband (50, 144 and 440MHz) antenna and see if it works out on the VHF marine band. Because VHF range is normally little more than line of sight a VHF antenna should be mounted as high as possible. With sailing boats the obvious and most popular site is the masthead. However, this is also the prime spot for navigation lights and weather instruments. When a VHF antenna is sited at the masthead it should be fixed to a bracket that places it a short distance from the navigation light to limit the angle of obstruction as shown in Fig 8.10.

CHAPTER 9

Mobile experimental activities

General antenna testing

I have found mobile radio to be a useful place to carry out experimental work on radio in general and antennas in particular. Some time ago I was asked to review a compact HF antenna. Because there was an antenna on my one and only mast I hit on the idea of using the car as a temporary antenna support. I also took the opportunity to test a magmount at the same time.

An antenna support, comprising four magmounts on a frame, is shown in Fig 9.1, which spreads the load over an area 380mm by 480mm (15 by 19in). The experimental arrangement set up to make antenna measurements is shown in Fig 9.2. The antenna is a commercial multiband Diamond antenna just to show the load a suitable magmount will support. When using large antennas with a mobile mount make sure that the vehicle body structure is strong enough for such an application.

A similar arrangement could be used by the serious mobile HF DXer to enable the use of a large efficient antenna (relative to the standard mobile antenna) from a fixed location.

Fig 9.1. Four small magmount magnets used with a frame to make a support for a larger structure. This magmount was made by Tennamast (see Appendix 1) and, at the time of writing, can be obtained from them on special order. *(Photo courtesy of* Practical Wireless *magazine)*

DDRR mobile roof rack antenna

I used a luggage rack as a platform for HF and VHF antennas a few times in the past – see Fig 5.6. I decided to see if the luggage rack itself could be used as an antenna.

The most likely candidate for the basis of the design was the DDRR antenna, described in [1]. The basic design is a short vertical section with the rest of a quarter-wavelength formed into a horizontal circular loop. No loading coils are necessary.

79

AMATEUR RADIO MOBILE HANDBOOK

Above: Fig 9.2. A Diamond antenna used as a mobile antenna, being tested in a static mobile arrangement and low wind conditions! This antenna uses tuned radials rather than the body of the vehicle as a counterpoise. *(Photo courtesy of* Practical Wireless *magazine)*

Right: Fig 9.3. Bird's-eye view of the DDRR roof rack antenna. The element overlap provides capacitance for tuning the antenna

This version of the DDRR antenna was constructed from 2mm copper tubing. Instead of using a circular horizontal section my design used a rectangle configuration, for two reasons:

First, a square or rectangle can be made up using straight sections of tubing, with the ends joined using 90° angle joints.

Second, a rectangular configuration looks more like a roof rack and, if constructed well enough, can even be used as a roof rack. The antenna is illustrated in Fig 9.3.

After reading the mathematical analysis of this antenna [2] by Robert Dome, W2WAM, I used a vertical height of 250mm (10in), which was as high as practicable if the antenna was to masquerade as a roof rack.

The antenna is fixed to the roof of the car by a bar roof rack – the sort used to transport ladders or timber, and is illustrated in Fig 9.4. As well as supporting the DDRR antenna, this rack also provides facilities to mount more conventional antennas.

The bars are set about 350mm (12in) apart and a sheet of aluminium is bolted to them. The roof rack support leg nearest to the vertical section of the antenna is removed to reduce RF absorption by the roof rack support.

Two pieces of Bakelite (or any other suitable insulating material) are fitted to the ends of the rack for mounting the DDRR element. The copper

CHAPTER 9: MOBILE EXPERIMENTAL ACTIVITIES

tubing elements are mounted to the insulated roof rack sections with plastic tube-to-wall fittings. Additionally I drilled the elements and used bolts and nuts to secure the elements, but not the variable capacitor section, to the insulated section. (This additional securing is only necessary if you require the antenna to double as an real roof rack.)

The feed end of the element is flattened so that it can be soldered to a short length of copper braid; the other end of the braid is attached to the door post with self-tapping screws to make a good electrical contact with the car metalwork – see Fig 9.5. This ground connection can be placed so that it is covered when the car door is closed. A thick piece of wire is also soldered to the ground end of the braid for a connection to the screen of the coaxial cable.

The antenna is shunt fed by tapping the feed-line centre up from the ground end of the element.

To keep the radiation resistance as high as possible the tuning capacitor value should only be large enough to tune the band. The capacitance is provided in this case by overlapping the top end of the element with the lowest part of the horizontal section. It is adjusted by sliding the top end of the element along the tube-to-wall fittings, thereby adjusting the overlap as shown in Fig 9.3. These fittings offer just enough friction to retain the element in the desired position. Calibration marks are used so that the overlap position can be set at any frequency within the tuneable range.

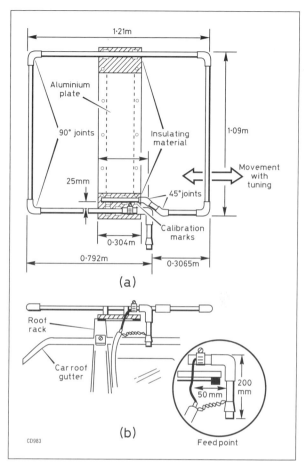

Fig 9.4. DDRR mobile roof rack antenna for 14MHz, top and side views

Transmitting antennas induce current into the metal body of a vehicle, as shown in Chapter 2. This current is related to current flow in the antenna, and that antenna current is indirectly proportional to the radiation resistance for a given transmitter power.

The DDRR antenna has a very low radiation resistance so the current induced into the vehicle body and vehicle wiring can be very high. While this is not a problem with older vehicles it could affect the running of modern ones using microprocessor-controlled engine management systems. A short antenna also produces very high voltages at the end – see Fig 9.6.

This antenna proved to be a very useful single-band antenna for 14MHz.

I didn't have to worry about removing it when parked in case it might get vandalised or stolen – after all it looks just like a luggage rack. Furthermore, I could drive into low-roofed multi-storey car parks without having to remove it.

This antenna was described in detail in *QST* November 1988 and *CQ Ham Radio* (Japan) December 1996.

Above: **Fig 9.5.** The feed end of the mobile DDRR antenna. The lower end of the element is connected to the bodywork of the car with tinned copper braiding, together with the braiding of the feeder coaxial cable. The centre of the cable is connected with a wire to an adjustable point (shunt feed) on the antenna element

Right: **Fig 9.6.** All lit up! The DDRR antenna produces a high voltage at the end – enough to illuminate a fluorescence tube at relatively low power. However, it is not a dangerous as it looks – the antenna is detuned if a hand is placed near it, causing the transceiver SWR power circuit to shut down the transmitter (although you are not advised to put it to the test!)

The toroidal antenna

The description of the antenna that follows is an example of how useful a mobile installation can be for evaluating antennas. What is described below is one of a series of antennas used to evaluate the toroid antenna proposed by G2AJV [3]. For a full description see reference [4]. It is not regarded as a serious solution to mobile antenna design although it can be used as one.

Construction

The antenna, designed for 14MHz, comprises two toroidal contra-wound inductances as shown in Fig 9.7.

Each toroid is constructed from 30 turns of 16SWG enamelled insulated wire. The outside diameter of each toroid is 150mm (6in) and the inside diameter 50mm (2in). Two sections of a plastic terminal block are used to support the ends of the toroids.

The toroids on the prototype are spaced 270mm apart and connected as shown in Fig 9.7. The whole structure is supported in a cylindrical cage made from white plastic garden fencing, using plastic tie-wraps as shown in Fig 9.8.

The base for the antenna was made from a sheet of aluminium (theoretically, the larger, the better) clamped to the roof of the car with magnetron magnets. This base fixing method also reduces the base/car roof capacitance. Eight holes are drilled in the base plate in groups of two. The metal between the holes is lifted so that tie-wraps can be inserted to fix the plastic cage to the base without touching the roof of the car.

CHAPTER 9: MOBILE EXPERIMENTAL ACTIVITIES

A hole is drilled for the earth point, which should be close to the antenna feedpoint when the antenna is assembled. The hole is countersunk so that the countersunk head of the earthing bolt is flush with the bottom of the base.

The outside edges of the base, facing the roof of the car, are faced with strips of plastic tape to prevent the base scratching the roof of the car. The base clamp magnets can be used with plastic sheet protection.

Performance

I made some comparison signal strength tests of the 14MHz toroid antenna using an 2.5m (8ft) home-made, centre-loaded vertical as a reference. This antenna had a lower section constructed from 22mm copper tubing and the air-spaced loading coil was 760mm (3in) in diameter. This reference antenna was fixed to the rear of the car, level with the bumper, with a good earth connection to the car chassis.

The toroid was fixed to the roof, as already described, and both antennas remained in place during the tests. I reasoned there would be little interaction because the unused antenna is detuned when the feeder is disconnected from the transceiver (load) when the antennas are changed over (the feeder was not a multiple of a quarter-wavelength at the test frequency).

Left: **Fig 9.7.** Diagram of the double-toroid antenna using series feed. C1 is used to resonate the antenna; C2 is used for matching

Above: **Fig 9.8.** A series-fed double-toroid antenna (14MHz) using home-made mobile roof mount. The clamping magnets are not shown. The large capacitor shown was a two-ganged 500pF variable used for finding the matching capacitance. Eventually the tuning was done with an air-spaced 50pF variable and a fixed 180pF Steafix fixed capacitor for loading

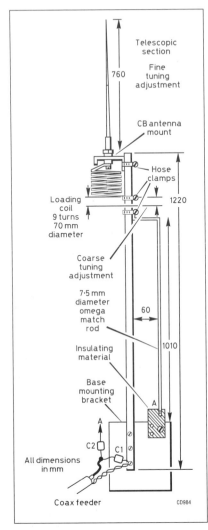

Fig 9.9. The omega-matched, centre-loaded antenna. Values for C1 and C2 were 200pF and 150pF respectively, obtained experimentally using variable capacitors. The coil is described in Chapter 4

Short-skip contacts to Europe were inconclusive; sometimes the vertical antenna outperformed the toroid and other times the toroid was the better performer.

I then tried ground-wave tests with GB3RS, the RSGB HQ station (at the time I was working at the RSGB). At distances of between a quarter and half a mile I transmitted a carrier which was carefully monitored on a power meter. John Crabbe, G3WFM, the senior station operator at the RSGB station, made measurements using the S-meter on the IC-781.

Most of the measurements gave the vertical a 0.5 to 1 S-point advantage. The exception was when the car was facing the HQ station, when the toroid antenna had the 1 S-point advantage.

The implications are that the toroid antenna may be particularly useful for low-band HF, particularly where space is at a premium.

An alternative feed for verticals

I constructed a centre-loaded vertical for 14MHz, which is described in Chapter 4 and illustrated in Fig 4.6. What was not described in Chapter 4 was the base insulator – there wasn't one.

Because I didn't have materials to hand to make a suitable base insulator for the 22mm (3/4in) lower section I considered fixing the antenna directly to the vehicle. Feeding the antenna could then be done with a gamma match, as is done with beam antennas. However, from the data I had at the time (traditional beam antenna book data) it seemed that the gamma rod length required would be too long to fit in the length of the antenna below the centre-loaded coil. I found that the matching rod length could be reduced using an omega match. The resultant antenna design is shown in Fig 9.9.

The omega match is constructed from aluminium tube with a small piece of insulating material to support the feedpoint and the omega matching capacitors.

The antenna was fixed directly to the metal bumper of the car with two thick copper braid straps to the main bodywork of the car to ensure the lowest possible resistance as shown in Fig 9.10.

I have learned more about gamma and omega matches since the design shown in Fig 9.9.

For example, if the matching rod were to be made from a smaller-diameter material, such as a length of hard-drawn 2mm copper wire, then the length of the matching rod would only have to be 0.853m (33.7in). Note the dimension given is for a gamma match rather than an omega match, thereby simplifying the matching network. Furthermore, three gamma matches (not omega matches) can be connected in parallel for a three-band mobile antenna for the 20, 15 and 10m bands.

I have tried a five-band antenna with five gamma matches in parallel but this arrangement did not work, presumably due to interaction.

Other experimental work

An amateur radio station fitted in a vehicle can be very useful for experimental work, particularly with antennas. In the early days of low frequency (73kHz) experiments, Mike Dennison, G3XDV, and I used vehicles when making comparative field-strength measurements. We used field-strength meters made from long-wave transistorised broadcast receiver components and communicated the results via UHF. I have also found it to be useful when trying out kite and balloon antennas – see Chapter 6.

A further use that I have put mobile radio to is the plotting of beam patterns for HF beams. The same could be done for VHF/UHF beams although I have not actually tried it.

Fig 9.10. The coaxial feed to the omega match. This was constructed in the days when cars had 'proper' bumpers made of metal! The support bracket is fixed to the metal bumper with a tinned copper braid earthing strip to the car chassis. The antenna is fixed to the mounting bracket with two nuts and bolts for easy removal

References

[1] *The ARRL Antenna Handbook*, 17th edn.
[2] *The Antenna Experimenter's Guide,* 2nd edn, Peter Dodd, G3LDO, RSGB, 1996.
[3] 'The G2AJV toroidal antenna', Roger Jennison, G2AJV, *Radio Communication* April and May 1994.
[4] 'Evaluation of the G2AJV toroidal antenna', Peter Dodd, G3LDO, *Radio Communication* August 1994.

CHAPTER 10

Walkabout mobile

When you operate /M using a VHF/UHF hand-held transceiver, it is good fun to take one of these diminutive little rigs on to a hill or mountain and see just how far you can work, particularly in 'lift' conditions. More recently amateur radio equipment manufacturers have brought out small transceivers so that HF operators can join in the fun. With HF it is possible to have intercontinental QSOs when conditions are right.

This chapter is devoted to what I describe as walkabout mobile – see Fig 10.1.

My definition of mobile manpack or handheld equipment, see Chapter 1, is that the transceiver or pack should be complete with its power supply and antenna, although it can be set down on a nearby wall or table when operated.

There is a type of operating known as *adventure radio* (mainly in the USA) where operators hike for miles along mountainous terrain, keeping in touch with tiny HF (usually homebrew) transceivers. Wire antennas are normally used which are fixed to a nearby tree. This is portable operation and is outside the scope of this chapter. The reason for this demarcation is that a full description of portable operation would require a book in its own right.

Fig 10.1. What is the best-dressed walkabout operator wearing? This HF pack uses an IC-705 manpack (or should it be womanpack?) with the TEK mobile antenna fitted with a 18MHz extension whip. The total weight of the pack is around 7kg

Fig 10.2. The complete G0CBM walkabout mobile installation in Portugal. It comprises four 300mm (12in) copper-plated steel tubular sections (ex-army whip) to make a 1.3m tall antenna. It uses a large capacity hat made from four 460mm (18in) telescopic whips. It has a centre-loading coil and a simple ATU at the base

Walkabout mobile is essentially QRP because of the power supply limitations; 5W RF output seems to be the best compromise between performance and time endurance.

The popularity of walkabout mobile has been rekindled by the availability of the FT-817 transceiver. The description of this radio is given later in the chapter and is based on a review [1] by Peter Hart, G3SJX.

However, it is possible to operate walkabout mobile with older, more-modest radios. An example of such an installation is the one used by G0CBM.

The G0CBM walkabout mobile

A few weeks ago I worked CT/G0CBM/M, who was on holiday in Portugal, on 18.16MHz from my mobile station. This station was using a manpack transceiver, running 6W from an FT-70G and using an integral home-made antenna and ATU. I had an interesting SSB QSO with the operator, Charles Wilkey. What follows is a description of his mobile station, which he sent to me on his return from his holiday.

The radio used in this installation is the rather unusual Yaesu FT-70G. This is a commercial radio with a continuous receiver and transmitter frequency coverage of 1 to 30MHz on modes LSB, USB, AM and CW. The frequency is selected by thumbwheels in 100Hz steps.

The power out is 10W although it has been set to 6W to reduce battery drain.

The dimensions of the FT-70 are 13in tall by 9in wide by 3in deep and it weighs 11kg with battery, so it is a fairly big radio compared with the FT-817.

There is a tendency to use small antennas with small radios, when what is really needed is the largest and most efficient antenna possible, particularly when operating QRP. G0CBM has designed a very good antenna system for his walkabout mobile. Theoretically, it will be more efficient than most vehicle mobile antennas and is worth a detailed description. The complete antenna is shown in Fig 10.2.

The antenna is designed so that it can be broken down and stored in a 330mm (13in) package (the height of the manpack with the battery fitted). This is shown in Fig 10.3.

The loading coils are made from silver-plated Airdux coils material with end plates of fibreglass PCB with the copper etched off. There are three separate coils (although only two are shown in Fig 10.3 and Fig 10.4). The coil dimensions are:

- Large coil for 80m: 76mm (3in) diameter, 60 turns
- Medium coil for 40m and 30m: 50mm (2in) diameter, 37 turns
- Small coil for 20, 17 and 15m: 50mm (2in) diameter, 11 turns

One of the most innovative aspects to this antenna is its built-in ATU. It comprises a series 100pF variable Jackson capacitor, which can be bypassed, and 330 and 180pF shunt capacitors, which can be switched in as required. This ATU is mounted in a die-cast box as shown in Fig 10.5. The PL519 plug connects the ATU to the transceiver antenna socket. The rest of the antenna is assembled, with the appropriate loading coil, on the end of the rod section at the top of the ATU.

The transceiver uses a counterpoise arrangement, comprising three 2.44m (8ft) PVC-covered lengths of wire. These counterpoise wires together with the body of the transceiver make a usable non-resonant arrangement that works well on

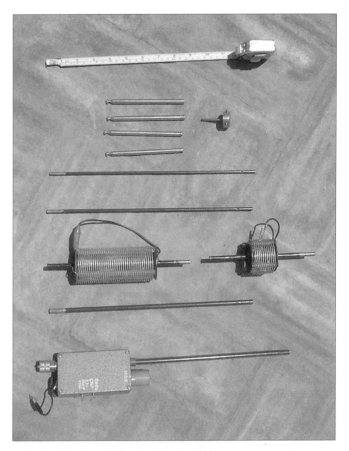

Above: Fig 10.3. Components of the G0CBM antenna. Top: capacity hat comprising four 460mm (18in) telescopic whips which screw into a brass boss. Four 300mm (12in) copper-plated steel tubular sections that make up the vertical section. Loading coils and ATU

Left: Fig 10.4. The loading coils. Left: the small coil for 20, 17 and 15m; right: the medium coil for 40m and 30m. There is also an 80m coil – not shown

AMATEUR RADIO MOBILE HANDBOOK

Fig 10.5. The ATU, with a series variable capacitor and fixed shunt capacitors. The fixed capacitors can be switched in with the lower switch and the variable capacitor shorted out with the upper switch

15m; it appears to avoids wild changes in impedance with frequency experienced with resonant counterpoise systems.

Matching the antenna to the transceiver in the field can be difficult but the following appears to work OK.

On 80m, plug in the appropriate coil. At the ATU select the 330pF shunt capacitor and variable series capacitor. Select tap on coil using crocodile clip and peak RF output by adjusting variable series capacitor.

On 40m, plug in the appropriate coil, select 180pF and variable series capacitor. Then adjust as described above.

On 20, 17 and 15m, plug in the appropriate coil, switch out shunt, and bypass series capacitors. Select tap with crocodile clip for maximum RF output.

Usually the station is packed into a rucksack with the minimum of extras, eg paper instead of logbook and A4 lists of country prefixes, but the weight is still 13kg. When required for operating it is then unpacked and set up on the ground, dry stone wall or rock.

The battery pack is a home-made arrangement using two 6.5Ah 6V gel cells. This enables the rig to be used for around eight hours of continuous QSO operating.

Counties worked from Portugal include VK, VE, WA, C56, P43, 9H3, J3 and Europeans.

It would appear best results are achieved when the antenna is clear of surrounding objects such as walls, buildings and especially trees.

The G3LDO walkabout mobile

I tried my hand at the construction of a walkabout mobile station. It was inspired by an earlier purchase of a 'knapsack' for an ex-army manpack radio from a local ex-government radio emporium in Worthing (we are lucky enough to still have one of these!).

The 'knapsack' comprises an aluminium radio equipment support plate with a waxed cloth cover. This cover has a number of flaps, which can be wrapped around the radio installation and fixed in place with Velcro fasteners. The trough section at the bottom of the aluminium plate was just the right size for the IC-706. Details of the installation are shown in Fig 10.6.

The installation was tested on 18MHz, my favourite mobile band – particularly at the weekend. This is a contest-free zone, with a much lower QRM level and excellent for testing QRP stations. The rig and power supply, using a 12V 6Ah, sealed lead-acid battery is just over 7kg. The ex-army radio knapsack is well designed with wide padded shoulder straps. The IC-706 is easy to operate by using the detachable front panel as shown in Fig 10.7.

The main difficulty with walkabout mobile is resonating the antenna.

With this type of installation the mass of metal normally found in a mobile installation, which is used as a counterpoise, is absent. I have experimented with various HF antenna arrangements and have come to the conclusion that G0CBM has got it right; although different ATU component values and counterpoise lengths may be required for smaller radios.

I tried my manpack with the Texas Bugcatcher antenna – see Fig 10.8. This worked surprisingly well on the lower-frequency bands where I expected more problems. I used a single counterpoise laid out along the ground and found that, provided the antenna was reasonably near resonance, the lowest SWR could be obtained just by altering the length of the counterpoise. In practical terms this was achieved by using a small roll of plastic insulated wire and unrolling the wire to obtain the correct counterpoise length. An arrangement using the case of a roll-up tape measure would probably be suitable.

Fig 10.6. The IC-706 mobile walkabout installation. The transceiver fits into a trough at the bottom of the aluminium support tray. The battery rests on a rubber or plastic cushion on the top of the transceiver and is held in place with a bungee strap. The antenna base is fixed to the top of the aluminium plate

FT-817 HF/VHF/UHF portable transceiver

The FT-817 measures only 135 by 38 by 165mm and weighs a little over 1kg, which represents a considerable weight and space saving compared with the two walkabout installations so far described. These factors have contributed to making this a very popular radio for walkabout mobile, and for these reasons it is described in some detail.

The FT-817 is supplied with a shoulder carrying strap, hand microphone (MH-31 as used on most Yaesu radios) and a three-piece 'rubber duck' style whip antenna for 6m/2m/70cm. There are two antenna sockets, a BNC on the front panel and SO-239 on the rear, and it is possible to select either front or rear separately for the four band groups HF/6m/2m/70cm. The rear-panel socket is used with the radio horizontal, for example on a table top, and the front-panel socket for a whip antenna with the radio carried vertically.

The radio can be powered using 9.6V to 13.8V, either from an external DC supply or from internal batteries. These batteries can be either eight

Fig 10.7. The IC-706 HF/VHF walk-mobile. The control panel is detached and can be operated very easily while walking. The control lead and the headphone/boom microphone leads are fixed to the shoulder strap with tape. The ex-army knapsack also has a pouch at the rear, which is used for stowing the control panel

AA-size alkaline cells or the Yaesu FNB-72 NiCd rechargeable battery pack which supplies 9.6V at 1000mAh capacity. Models supplied from UK dealers include the rechargeable battery as standard, together with a mains wall charger. Although the manual states that this charger can only be used when the radio is switched off, it is rated at 500mA, which is just sufficient to power the radio on receive as well as charging the battery, although insufficient to provide power on transmit. A higher-power external 13.8V supply (rated at 2.5A) will allow the batteries to be charged while also using the radio to the full. The charging time can be set to 6, 8 or 10 hours and the remaining time to full charge is displayed when the radio is switched off. This is reset if the charging current is interrupted for any reason.

The receiver in the FT-817 tunes from 100kHz to 56MHz, 76 to 108MHz (wide-band FM mode only), 108 to 154MHz and 420 to 470MHz. The transmitter is enabled only within the exact amateur allocations with variants for different regions. Up/down keys scroll through the various amateur bands, general coverage and broadcast bands, and another pair of up/down keys scrolls through the modes – LSB, USB, CW, CW-R, AM, FM, digital and packet. Digital mode uses AFSK on SSB modes and is intended for RTTY, PSK31, SSTV etc. Packet mode uses FM and has settings for both 1200 and 9600 baud operation.

There are four transmit power output settings – 5W, 2.5W, 1W and 0.5W with 2.5W as the default setting on internal batteries.

The radio is solidly constructed on a die-cast aluminium chassis with controls on the front and top edge, a 55mm diameter speaker in the top, microphone and headphone sockets on the side, access to the batteries underneath and sockets on the rear for key, data terminals and PC control. The radio is fully controllable from a PC but needs a special interface cable available as an option which includes a built-in RS232 level converter. Electrically, the radio uses a double-conversion superhet receiver with IFs of 68.33MHz and 455kHz. The transmitter PA, PA drivers and receiver front-end mixer are all wide band, covering a remarkably wide frequency range from 160m to 70cm.

CHAPTER 10: WALKABOUT MOBILE

Principal features

The FT-817 is packed with features – indeed virtually the full feature set as found on most larger radios is provided. It is always a challenge with a small radio and hence limited panel area to provide a simple and user-friendly access to its many functions. Some dedicated controls are essential, such as tuning, band and mode change, volume etc, but other functions are accessible through context and menus. The front panel is shown in Fig 10.9.

Three buttons below the display select most of the functions of the radio. A quick press of the 'F' key displays the function associated with these buttons and a small click-step rotary control 'Select' scrolls through 12 sets of button allocations. In addition the menu system allows some 57 parameters of the radio to be set. This is accessed also with the 'F' key and the 'Select' control with the rotary tuning control to set the parameter.

Tuning makes use of a small rotary control in conjunction with the detented 'Select' control mentioned above. Tuning is in 10Hz steps at 2kHz per revolution or 20Hz steps at 4kHz per revolution on SSB/CW which is rather slow and tedious with the small 25mm diameter knob and so the 'Select' control which tunes in 1, 2.5 or 5kHz steps is used for coarse navigation. This also provides 1MHz stepping for large frequency excursions. AM or FM tuning is normally achieved via the 'Select' control, with a selection of separate mode-dependent step sizes, although rotary control tuning at 10 times the SSB rates can be selected for this purpose.

Fig 10.8. Test set-up using the IC-705 manpack with the Texas Bugcatcher antenna. The roll of wire is a makeshift counterpoise, which worked very well on the lower frequency bands

Fig 10.9. The FT-817 front panel. The controls are described in the text

Despite its compact size, the FT-817 provides comprehensive memory features. 200 regular memories are included which may be partitioned into 10 groups of 20 channels and each channel may have an eight-character alpha-numeric label attached for easy identification. A one-touch quick memory store allows one frequency to be rapidly stored and recalled, and a separate home channel for each of the four band groups may be selected at the push of a button.

The radio includes a number of power-saving features. Auto power-off will automatically turn off the radio if there has been no control activity for a period (one to six hours) and the transmit time can be limited (one to 20 minutes). The display backlighting can be turned on or off or set to auto (default) where the backlighting is on for only five seconds after any key presses. The backlighting colour can be set to blue or orange – G3SJX preferred orange in most situations. The LCD indicates frequency to 10Hz resolution, memory channels or labels, mode and VFO status, and a number of small icons. The battery voltage can be permanently displayed and there is a bargraph-type S-meter which indicates power, SWR, ALC level or modulation on transmit. One of the menu settings shows DSP as a label for one of the buttons. Don't be misled – this selects double display height for clearer frequency indication – the radio is not fitted with digital signal processing.

Mainly HF features

Twin VFOs are incorporated, each with separate band stores. These can be used separately for CW and SSB segments or used together for split-frequency operation. A clarifier (receiver incremental tuning) covers ±10kHz and functions on receive only, IF shift helps in reducing adjacent channel interference and an IF noise blanker is included for reduction of ignition and other impulse noise. The radio is provided with a 2.4kHz ceramic IF filter for SSB and CW modes but space is provided to install a 10-pole Collins mechanical filter, either a 500Hz filter for CW and digital modes or a 2.3kHz filter with improved shape factor for SSB.

Other receive features include fast/slow AGC, RF gain control/squelch and variable CW pitch over the range 300 to 1000Hz. For strong-signal situations, the receive preamp may be switched out (IPO) and a 10dB attenuator may also be switched in. On 2m and 70cm the receive preamp is permanently in circuit.

VOX is provided functioning on all voice modes but there is no speech processor. A semi-break-in system is included for CW with recovery delay times separately adjustable for CW and VOX. Although not specifically designed for full break-in, the minimum recovery delay time of 10ms effectively emulates QSK operation. A built-in CW electronic keyer is adjustable in speed over the range 4 to 60WPM and has adjustable dot/dash weighting, but does not include any memories or contest-related features. For occasional or emergency use it is possible to assign the up/down keys on the microphone for generation of dots and dashes.

The FT-817 is well equipped with facilities to handle digital and packet modes. Audio input levels are separately adjustable for each data mode as are display and pass-band offsets. As well as the predefined modes of PSK31, RTTY and packet, two user-definable modes (USB and LSB) are also included. These can be used for SSTV or a future new digital mode. The FT-817 with a small laptop PC makes an effective and very lightweight station for PSK31, given the excellent low power performance of that mode.

Mainly VHF/UHF features

The FT-817 includes all the features which are available on a modern FM hand portable. Both wide and narrow FM modes are provided, covering 25/12.5kHz channelling on VHF/UHF or 10kHz on 29/50MHz. Both the receiver bandwidth and transmitter deviation levels are set appropriately.

For repeater operation, the shift is separately programmable on 10m, 6m, 2m and 70cm and can be automatically selected according to the band plan in use in the relevant region on 2m and 70cm. The transmit and receive frequencies can be reversed by a single key press to check for activity on a repeater input channel. Both a 1750Hz tone burst and a CTCSS tone encoder are provided for repeater access and a CTCSS decoder provides tone search to detect and store the CTCSS tone transmitted by a received station or repeater.

A digital code squelch (DCS) system is also built-in. This uses one of 104 selectable codes to implement a squelch-controlled link and is more robust and less prone to false triggering than CTCSS. A code search feature allows the DCS code transmitted by a received station to be detected and stored. Complementary to the DCS system is the ARTS (auto range transponder system) also fitted. This uses DCS signalling to inform when you and another ARTS-equipped station are within communications range.

A number of scanning related features are provided. Scanning can be initiated in VFO mode, up or down from any start frequency or between programmed limits with user-programmable pause/resume status. In memory mode, memory channels can be scanned sequentially up or down and channels can be selected for skipping. 'Dual watch' allows VFO-B to be checked every five seconds whilst using VFO-A for normal communication purposes. In a similar way, 'priority channel checking' lets you operate on a memory channel while checking memory channel 1 every five seconds. 'Smart search' is a useful feature when travelling in a new area and functions on AM and FM. A scan is initiated in VFO mode and the first 50 active channels are loaded into special memory.

The FT-817 also includes a spectrum scope monitor which monitors activity on five channels either side of the receive frequency and displays relative signal strength as a bargraph on the LCD. Normal receiver operation is disabled whilst the spectrum monitor is functioning. Although operational on all modes, the result is only really meaningful for monitoring FM channels. The IF bandwidth for the spectrum scan is set to the FM bandwidth and channels are scanned according to the step size set for the 'Select' channel stepper. This step size needs to be set appropriately to get the desired result, normally the operational channel step size.

On the air

The receiver in the FT-817 really performed very well. G3SJX notes that in these tests on his home station antennas, there were very few signals which couldn't be copied as well on the FT-817 as on his FT-1000MP. On the LF bands with large antennas the preamp needed to be switched out (IPO) in most cases to avoid overload but rarely was it necessary to switch in the attenuator. The receive audio was fairly toppy but good communications quality and with plenty of punch. The AGC slow setting was used on all modes; in the fast setting background noise would return to full level between Morse characters and speech symbols in a disconcerting way. Surprisingly the S-meter decay was much slower to respond. The filters are good and the narrow CW filter well recommended. Broadcast AM and wide-band FM both gave excellent results and quality. The transmit audio was clear and punchy with good quality and the CW break-in system was effective. G3SJX worked a number of DX stations with remarkable ease.

The control ergonomics for most functions are quite cleverly arranged and easily mastered after a brief learning period, although it would be preferable to have the button legends to be displayed continuously. These share the same display area as the S-meter and revert back to the S-meter display a second or so after each key press. Although there is a low-battery icon, it is not very attention grabbing. When the battery voltage drops, a point is reached when the radio just switches off with no prior warning.

Some optional extras

A range of base-loaded telescopic whip antennas intended for use with the FT-817 is available (see Appendix 1), and these are shown in Fig 10.10. They plug into the BNC connector on the front panel. Each comprises a 4ft telescopic whip section in conjunction with a loading inductor moulded into the base. The AT series are monoband antennas with separate models covering all bands from 80m to 70cm. Fully extended, the length is 1.4m, collapsing down to about 26cm. The ATX-Walkabout is a novel multiband antenna with a tapped loading inductor and a jumper lead which shorts out various sections of the inductor. This single antenna is adjustable on all bands from 80m to 6m, is 1.65m long fully extended and only 32cm long when dismantled.

The antennas are tuned by adjusting the length of the telescopic section whilst observing the VSWR display on the FT-817. As the antennas are very short compared with the operating wavelength, the bandwidth is quite narrow and tuning is fairly sharp, particularly on the lower-frequency bands. A ground-plane wire or earth lead must be connected to

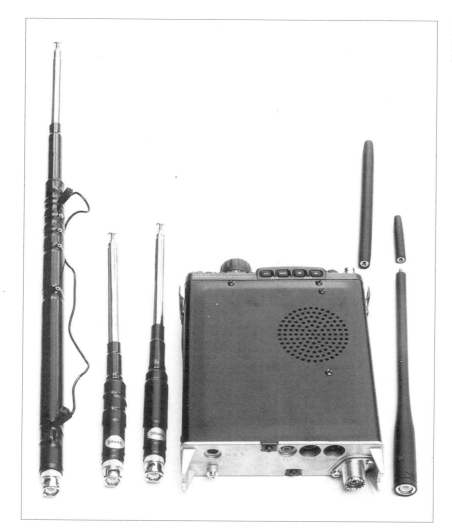

Fig 10.10. FT-817 with ATX-Walkabout and AT single-band antennas

the ground terminal on the back of the radio to obtain any reasonable performance on receive and is a must on transmit. As these are fairly rigid antennas, care should be taken to avoid any knocks or undue strain which may damage the BNC socket. 'Rubber duck' VHF antennas are flexible and present much less strain on the antenna socket.

Reference

[1] 'The Yaesu FT-817 review', Peter Hart, G3SJX, *Radcom* June 2001.

APPENDIX 1

Resources

Antenna and equipment suppliers

Haydon Communications
This company markets the Q-TEK MA5M HF antenna, which is used with the bike and walkabout mobile described in this book. Also available is a range of single, dual and triband Q-TEK antennas from 6m to 70cm. Mail order tel: 01708 862524.

Moonraker
Specialises in antenna equipment and markets the Ampro antennas. These are single-band antennas from 6m to 160m. There is also the Ampro MB5, which covers 10/15/20/40/80m, together with a range of single, dual and triband antennas from 6m to 70cm. This company also sells various magmounts and hatchback mounts.

All the products described are illustrated on their web site: www.amateurantennas.com. Their address is: Unit 12, Cranfield Road Units, Cranfield Road, Woburn Sands, Bucks, MK17 8UR. Tel: 01908 281705.

Nevada
Sells the Comet range of mobile antennas, which includes CA-UHV. This multiband antenna covers 7, (14), 21, 28, 50, 144 and 430MHz. An additional coil is required for 14MHz.

Also available is a range of single, dual and triband antennas from 6m to 70cm, the Outbacker range and the low-profile 'flexible' antennas for the Yaesu FT-817.

This company sells various magmounts and hatchback mounts. In addition they have a wide range of HF and VHF/UHF transceivers and HF ATUs, including a used equipment range.

All the products described are illustrated on their web site: www.nevada.co.uk. Their address is: Unit 1, Fitzherbert, Spur, Farington, Portsmouth, PO6 1TT. Tel: 01908 281705.

Tennamast
The multi-magmount shown in Fig 9.1, comprising four small magmount magnets used with a frame, is made by Tennamast on special order. The

address is: Tennamast Scotland Ltd, 81 Mains Road, Beith, Ayrshire, KA15 2HT. Tel: 01505 503824.

Waters & Stanton
This company has, for the last three years, produced a very comprehensive colour catalogue, which gives some idea of the range of equipment available for the radio amateur who is interested in mobile operating.

The most versatile and effective HF antenna is probably the remote tune WBB-3. Also available is the Texas Bugcatcher; both these antennas are described in this book.

In addition there is the ATX-Walkabout, a novel multiband antenna with a tapped loading inductor and a jumper lead which shorts out various sections of the inductor. This single antenna is adjustable on all bands from 80m to 6m. It is 1.65m fully extended and only 32cm dismantled and is intended for use with the FT-817.

W&S sell the air-spaced coil stock suitable for constructing centre-loading coils.

In addition they market a range of single, dual and triband antennas from 6m to 70cm, magmounts, hatch mounts and cable kits. They sell a full range of radio equipment, including the FT-817, and specialist equipment such as the Kenwood TH-D7E and the Garmin Street Pilot GPS, as used with APRS. The list is too long to mention here – get the catalogue.

Waters & Stanton, 22 Main Road, Hockley, Essex, SS5 4QS. Freephone order tel: 08000 73 7388. General tel: 01702 206835/204965. Web: www.wsplc.com.

Miscellaneous items
APRS
G0TRT's web page, covering APRS basics, including digipeating and unproto path presentation: http://go.to/APRSUK.

IGATEs in the UK by Keith, G6NHU: www.mb7uiv.co.uk.

The Bob Bruninga, WB4APR, homepage: http://web.usna.navy.mil/~bruninga/aprs.html.

UI-View written by Roger Barker, G4IDE. His homepage is: www.packetradio.org.uk.

APRS software, but there is software available also for DOS, Windows®, WinCE®, Linux®, Macintosh®, PalmOS® and probably more (see G0TRT's web site).

APRS Tracks, Maps & Automobiles by Stan Horzepa, WA1LOU, available from the RSGB Bookshop.

Boat rigging insulators
Some (not all) ship chandlers sell rigging insulators. The examples shown in the book were obtained from: Arun Canvas and Rigging, The Shipyard, Rope Walk, Littlehampton, West Sussex, BN17 5DG. Tel: 01903 732561.

APPENDIX 1: RESOURCES

Kites

Kites can be obtained from any good kite shop; most towns have one. For example, my nearest supplier is Carousel Kites, 23 New Broadway, Tarring Road, Worthing. Tel: 01903 212002.

The kite I use is the Delta Supreme. This kit is excellent for supporting antennas and is 200 by 110cm with fibreglass frame, and comes with line on halo reel. It can be obtained from Kite Corner, 675 Watford Way, London, NW7 3JR. Tel: 020 895 90619.

Also available is a book, *Kite Cookery*, by Squadron Leader Don Dunford, MBE, and obtainable from Kite Corner – see above.

Magnetic material

The magnetic strip material used for making the mobile ground coupler, see Fig 5.9, can be obtained from many sign writing companies. In my case the material was obtained from: Hilan plc, Hilan House, Clinton Place, Seaford, East Sussex, BN25 1NP. Tel: 01870 3333 244.

Noise suppression components

Alternator suppression capacitors have special fittings for fixing to the side of the alternator. Electric motor noise filters, with a current rating of up to 7.5A, can be fitted in the motor leads of windscreen wipers, heater blowers or fuel pumps. Any of these items can be found at your local branch of Halfords or any vehicle parts outlet.

The Icom OPC-639 power lead filter is available from Waters & Stanton.

APPENDIX 2

Voice repeater lists

Voice repeater stations, together with maps showing most of their locations. The lists show the current repeater situation as of September 2001 – links to up-to-date lists are at www.rsgb.org.

Callsign	Locator	Old chan	New chan	Rx freq (MHz)	Tx freq (MHz)	Site	CTCSS	Keeper
GB3CJ1	IO92NF	R29		29.540	29.640	Northampton	No	G4SCJ
GB3CJ2	IO92NG	R23		29.540	29.640	Northampton	No	G4SCJ
GB3EF	JO02PB	R50-1	RF72	51.220	50.720	Ipswich	110.9	G0VDE
GB3UM	IO92IQ	R50-3	RF74	51.240	50.740	Leicester	77	G8OBP
GB3HF	JO00HV	R50-5	RF76	51.260	50.760	Hastings	No	G1DVU
GB3UK	IO83RO	R50-6	RF77	51.270	50.770	Bolton	82.5	G8NSS
GB3PX	IO92XA	R50-7	RF78	51.280	50.780	Hertfordshire	77	G4NBS
GB3SX	IO83WA	R50-8	RF79	51.290	50.790	Stoke on Trent	103.5	G8DZJ
GB3HX	IO93BP	R50-9	RF80	51.300	50.800	Huddersfield	82.5	G0PRF
GB3FX	IO91OF	R50-10	RF81	51.310	50.810	Surrey	82.5	G4EPX
GB3RR	IO93JA	R50-11	RF82	51.320	50.820	Nottingham	71.9	G4TSN
GB3WX	IO81UD	R50-12	RF83	51.330	50.830	Wincanton	77	G3ZXX
GB3AM	IO91QP	R50-13	RF84	51.340	50.840	Amersham	77	G0RDI
GB3PD	IO90KT	R50-14	RF85	51.350	50.850	Portsmouth	71.9	G4JXL
GB3AE	IO71PR	R50-15	RF86	51.360	50.860	Tenby	94.8	GW0WBQ
GB3BY	IO82UI	R50-15	RF86	51.360	50.860	Kidderminster	67.1	G8EPR
GB3AS	IO84KS	R0	RV48	145.000	145.600	Carlisle	No	G4KFN
GB3CF	IO92IQ	R0	RV48	145.000	145.600	Leicester	No	M0BKH
GB3EL	JO01AM	R0	RV48	145.000	145.600	East London	No	G4RZZ
GB3FF	IO86JB	R0	RV48	145.000	145.600	Burntisland	No	MM0AMV
GB3LY	IO65NC	R0	RV48	145.000	145.600	Limavady	No	GI3USS
GB3MB	IO83UO	R0	RV48	145.000	145.600	Bury	No	G8NSS
GB3SR	IO90WT	R0	RV48	145.000	145.600	Brighton	No	G4PAP
GB3SS	IO87KM	R0	RV48	145.000	145.600	Moray	No	GM7LSI
GB3WR	IO81PH	R0	RV48	145.000	145.600	Wells	94.8	G0TJP
GB3YC	IO94SC	R0	RV48	145.000	145.600	Driffield	No	G0OII
GB3CQ	IO92PM	R0X	RV49	145.0125	145.6125	Corby	No	G1DIW
GB3GD	IO74SG	R1	RV50	145.025	145.625	Douglas	No	GD3LSF
GB3HG	IO94KI	R1	RV50	145.025	145.625	Northallerton	No	G0RHI
GB3KS	JO01PD	R1	RV50	145.025	145.625	Dover	103.5	G4OJG
GB3NB	JO02NM	R1	RV50	145.025	145.625	Norwich	No	G8VLL

AMATEUR RADIO MOBILE HANDBOOK

Callsign	Locator	Old chan	New chan	Rx freq (MHz)	Tx freq (MHz)	Site	CTCSS	Keeper
GB3NG	IO87XO	R1	RV50	145.025	145.625	Fraserburgh	No	MM1CAC
GB3NW	IO82VE	R1	RV50	145.025	145.625	Worcester	67.1	G4IDF
GB3PA	IO75QV	R1	RV50	145.025	145.625	Renfrewshire	No	GM7OAW
GB3SC	IO90BR	R1	RV50	145.025	145.625	Bournemouth	No	G0API
GB3SI	IO70GE	R1	RV50	145.025	145.625	St Ives	No	G3NPB
GB3WL	IO91UM	R1	RV50	145.025	145.625	West London	No	G8SUG
GB3AY	IO75OR	R2	RV52	145.050	145.650	Ayrshire	No	GM0WUX
GB3BF	IO92SD	R2	RV52	145.050	145.650	Bedford	77	G8MGP
GB3EC	IO92DK	R2	RV52	145.050	145.650	Birmingham	67.1	G4KQV
GB3GJ	IN89WE	R2	RV52	145.050	145.650	St Hellier	No	GJ0NSG
GB3HS	IO93RS	R2	RV52	145.050	145.650	Hull	88.5	G7JZD
GB3MN	IO83XH	R2	RV52	145.050	145.650	Stockport	No	G8LZO
GB3OC	IO88LX	R2	RV52	145.050	145.650	Kirkwall	No	GM0HQG
GB3PO	JO02NB	R2	RV52	145.050	145.650	Ipswich	No	G8CPH
GB3SB	IO85VN	R2	RV52	145.050	145.650	Selkirk	No	GM0FTJ
GB3SL	IO91XK	R2	RV52	145.050	145.650	Crystal Palace	No	G4PEB
GB3TR	IO80FM	R2	RV52	145.050	145.650	Torquay	No	G8XST
GB3WH	IO91EM	R2	RV52	145.050	145.650	Swindon	118.8	G4LDL
GB3DW	IO72VW	R2X	RV53	145.0625	145.6625	Criccieth	110.9	GW4KAZ
GB3SH	IO90HW	R2X	RV53	145.0625	145.6625	Cantell School	No	M1AFM
GB3BX	IO82XP	R3	RV54	145.075	145.675	Wolverhampton	67.1	G4JLI
GB3ES	JO00HV	R3	RV54	145.075	145.675	Hastings	103.5	G7LEL
GB3LD	IO84KE	R3	RV54	145.075	145.675	Dalton, Cumbria	No	G7MCE
GB3LG	IO76HD	R3	RV54	145.075	145.675	Lochgilphead	No	GM4WMM
GB3LU	IP90KD	R3	RV54	145.075	145.675	Lerwick	No	GM4SWU
GB3NA	IO93GN	R3	RV54	145.075	145.675	Barnsley	71.9	G4LUE
GB3PE	IO92XO	R3	RV54	145.075	145.675	Peterborough	94.8	G1ARV
GB3PR	IO86GI	R3	RV54	145.075	145.675	Perth	No	GM8KPH
GB3RD	IO91JM	R3	RV54	145.075	145.675	Reading	118.8	G8DOR
GB3SA	IO81AO	R3	RV54	145.075	145.675	Swansea	No	GW4JGU
GB3WZ	IO83MB	R3	RV54	145.075	145.675	Wrexham	110.9	GW7TKZ
GB3AR	IO73VC	R4	RV56	145.100	145.700	Caernarfon	No	GW4KAZ
GB3BB	IO81OE	R4	RV56	145.100	145.700	Brecon	No	GW0ABT
GB3BT	IO85XT	R4	RV56	145.100	145.700	Berwick on Tweed	No	GM1JFF
GB3EV	IO84SQ	R4	RV56	145.100	145.700	Cumbria	No	G0IYQ
GB3HH	IO93BF	R4	RV56	145.100	145.700	Buxton	No	G4IHO
GB3HI	IO76DL	R4	RV56	145.100	145.700	Oban	No	GM1YUO
GB3KN	JO01GH	R4	RV56	145.100	145.700	Maidstone	103.5	G3YCN
GB3VA	IO91LT	R4	RV56	145.100	145.700	Aylesbury	No	G8BQH
GB3WD	IO70XN	R4	RV56	145.100	145.700	Plymouth	No	G6URM
GB3KY	JO02FS	R4X	RV57	145.1125	145.7125	Norfolk	94.8	G1SCQ
GB3AG	IO86ON	R5	RV58	145.125	145.725	Forfar	94.8	GM1CMF
GB3BI	IO77WO	R5	RV58	145.125	145.725	Inverness	67.1	GM0JFK
GB3CG	IO81VU	R5	RV58	145.125	145.725	Gloucester	No	G6AWT
GB3DA	JO01GR	R5	RV58	145.125	145.725	Chelmsford	110.9	G6JYB
GB3LM	IO93RF	R5	RV58	145.125	145.725	Lincoln	No	G8VGF
GB3NC	IO70OI	R5	RV58	145.125	145.725	St Austell	No	G3IGV
GB3NI	IO74CO	R5	RV58	145.125	145.725	Belfast	No	GI3USS
GB3SN	IO91LC	R5	RV58	145.125	145.725	Alton	No	G4EPX
GB3TP	IO93BV	R5	RV58	145.125	145.725	Keighley	No	G7HEN
GB3TW	IO94DT	R5	RV58	145.125	145.725	Durham	No	G8YWK
GB3VT	IO83WA	R5	RV58	145.125	145.725	Stoke on Trent	No	G8DZJ
GB3AL	IO91QP	R5X	RV59	145.1375	145.7375	Amersham	77	G0RDI

APPENDIX 2: VOICE REPEATER LISTS

Callsign	Locator	Old chan	New chan	Rx freq (MHz)	Tx freq (MHz)	Site	CTCSS	Keeper
GB3ZA	IO82PB	R5X	RV59	145.1375	145.7375	Hereford	118.8	G0JWJ
GB3BC	IO81KP	R6	RV60	145.150	145.750	Newport, Gwent	94.8	GW8ERA
GB3CS	IO75XX	R6	RV60	145.150	145.750	Lanarkshire	No	GM4COX
GB3MP	IO83IF	R6	RV60	145.150	145.750	Denbigh	No	G7OBW
GB3MX	IO93JD	R6	RV60	145.150	145.750	Mansfield	71.9	G0UYQ
GB3PI	IO92XA	R6	RV60	145.150	145.750	Royston	77	G4NBS
GB3WS	IO91WB	R6	RV60	145.150	145.750	Crawley	No	G4EFO
GB3NE	IO91HJ	R6X	RV61	145.1625	145.7625	Berkshire	118.8	G8JIP
GB3DG	IO74UV	R7	RV62	145.175	145.775	Gatehouse of Fleet	No	GM4VIR
GB3FR	JO03AE	R7	RV62	145.175	145.775	Spilsby	71.9	G8SFU
GB3GN	IO87SC	R7	RV62	145.175	145.775	Aberdeen	No	GM1XEA
GB3IG	IO68QE	R7	RV62	145.175	145.775	Stornoway	No	GM0PWS
GB3NL	IO91XP	R7	RV62	145.175	145.775	Enfield	No	G3TZZ
GB3PC	IO90KT	R7	RV62	145.175	145.775	Portsmouth	No	G4NAO
GB3PW	IO82HL	R7	RV62	145.175	145.775	Newtown, Powys	No	GW4NQJ
GB3RF	IO83US	R7	RV62	145.175	145.775	Burnley	No	G0DFO
GB3TE	JO01OT	R7	RV62	145.175	145.775	Clacton-on-Sea	103.5	G7HJK
GB3WK	IO92FH	R7	RV62	145.175	145.775	Leamington	No	G6FEO
GB3WT	IO64JQ	R7	RV62	145.175	145.775	Omagh	No	GI3NVW
GB3WW	IO71XT	R7	RV62	145.175	145.775	Cross Hands, Dyfed	No	GW6ZUS
GB3SF	IO93BF	R7X	RV63	145.185	145.785	Buxton	No	G4IHO
GB3BN	IO91OJ	RB0	RU240	434.600	433.000	Bracknell	No	G4HLF
GB3CK	JO01CP	RB0	RU240	434.600	433.000	Ashford	103.5	G6ZAA
GB3DT	IO80WU	RB0	RU240	434.600	433.000	Blandford Forum	No	G8BXQ
GB3EX	IO80FP	RB0	RU240	434.600	433.000	Exeter	No	G8UWE
GB3LL	IO83BH	RB0	RU240	434.600	433.000	Llandudno	No	GW8WFS
GB3MK	IO92OB	RB0	RU240	434.600	433.000	Milton Keynes	No	M5AET
GB3NR	JO02PP	RB0	RU240	434.600	433.000	Norwich	No	G8VLL
GB3NT	IO94FW	RB0	RU240	434.600	433.000	Newcastle upon Tyne	No	G8YWK
GB3NY	IO94TG	RB0	RU240	434.600	433.000	Scarborough	No	G4EEV
GB3PF	IO83SS	RB0	RU240	434.600	433.000	Blackburn	No	G0DFO
GB3PU	IO86GI	RB0	RU240	434.600	433.000	Perth	No	GM8KPH
GB3SO	IO92XX	RB0	RU240	434.600	433.000	Boston	71.9	G8SFU
GB3SV	JO01BU	RB0	RU240	434.600	433.000	Bishops Stortford	No	G1NOL
GB3US	IO93GJ	RB0	RU240	434.600	433.000	Sheffield	103.5	G3RKL
GB3WN	IO82XP	RB0	RU240	434.600	433.000	Wolverhampton	67.1	G4OKE
GB3BA	IO87SC	RB1	RU242	434.625	433.025	Aberdeen	No	GM1XEA
GB3BV	IO91SR	RB1	RU242	434.625	433.025	Hemel Hempstead	No	G8BQH
GB3DV	IO93JK	RB1	RU242	434.625	433.025	Doncaster	71.9	G4LUE
GB3EM	IO92OT	RB1	RU242	434.625	433.025	Waltham	No	G8WWJ
GB3HJ	IO93FX	RB1	RU242	434.625	433.025	Harrogate	118.8	G3XWH
GB3HO	IO91UC	RB1	RU242	434.625	433.025	Horsham	No	G7JRV
GB3MA	IO83UO	RB1	RU242	434.625	433.025	Bury	No	G8NSS
GB3TC	IO80SX	RB1	RU242	434.625	433.025	Wincanton	No	G3OOL
GB3WA	IO81UD	RB1	RU242	434.625	433.025	Great Ground Farm	No	G3ZXX
GB3AV	IO91OT	RB2	RU244	434.650	433.050	Aylesbury	No	G8BQH
GB3CH	IO70SM	RB2	RU244	434.650	433.050	Minions	No	G1RXR
GB3CI	IO92PM	RB2	RU244	434.650	433.050	Corby	No	G8MLA
GB3EK	JO01QJ	RB2	RU244	434.650	433.050	Margate	103.5	G4TKR
GB3FC	IO83LU	RB2	RU244	434.650	433.050	Blackpool	82.5	G6AOS
GB3HK	IO85ON	RB2	RU244	434.650	433.050	Selkirk	No	GM0FTJ
GB3LS	IO93RF	RB2	RU244	434.650	433.050	Lincoln	No	G8VGF

Callsign	Locator	Old chan	New chan	Rx freq (MHz)	Tx freq (MHz)	Site	CTCSS	Keeper
GB3LV	IO91XP	RB2	RU244	434.650	433.050	Enfield	No	G3TZZ
GB3NN	JO02JV	RB2	RU244	434.650	433.050	Wells Next The Sea	No	G0FVF
GB3NX	IO91XC	RB2	RU244	434.650	433.050	Crawley	No	G0DSU
GB3OS	IO82WL	RB2	RU244	434.650	433.050	Stourbridge	No	G1PKZ
GB3PH	IO90LU	RB2	RU244	434.650	433.050	Portsmouth	No	G8PGF
GB3ST	IO83WA	RB2	RU244	434.650	433.050	Stoke on Trent	No	G8DZJ
GB3UL	IO74CO	RB2	RU244	434.650	433.050	Belfast	No	GI3USS
GB3YS	IO80QX	RB2	RU244	434.650	433.050	Yeovil	No	G3UGR
GB3CC	IO90OU	RB3	RU246	434.675	433.075	Chichester	88.5	G3UEQ
GB3ER	JO01GR	RB3	RU246	434.675	433.075	Chelmsford	110.9	G6JYB
GB3HL	IO91UM	RB3	RU246	434.675	433.075	West London	No	G8SUG
GB3HU	IO93RS	RB3	RU246	434.675	433.075	Weedley Farm	No	G3TEU
GB3KA	IO75UO	RB3	RU246	434.675	433.075	Kilmarnock	No	GM0WUX
GB3KR	IO82VJ	RB3	RU246	434.675	433.075	Kidderminster	No	G8NTU
GB3MD	IO93JD	RB3	RU246	434.675	433.075	Mansfield	71.9	G0UYQ
GB3NH	IO92NF	RB3	RU246	434.675	433.075	Northampton	77	G4IIO
GB3TD	IO91DL	RB3	RU246	434.675	433.075	Swindon	No	G4XUT
GB3VS	IO80LX	RB3	RU246	434.675	433.075	Taunton	No	G4UVZ
GB3GC	IO93NQ	RB4	RU248	434.700	433.100	Goole	No	G0GLZ
GB3IH	JO02OB	RB4	RU248	434.700	433.100	Ipswich	No	G8CPH
GB3IW	IO90JO	RB4	RU248	434.700	433.100	IOW	71.9	G1VGM
GB3KL	JO02FS	RB4	RU248	434.700	433.100	Kings Lynn	No	M1ANH
GB3LE	IO92IQ	RB4	RU248	434.700	433.100	Leicester	No	M0BKH
GB3NK	JO01DH	RB4	RU248	434.700	433.100	Wrotham	103.5	G8JNZ
GB3OH	IO85EX	RB4	RU248	434.700	433.100	Bo'ness	No	GM6WQH
GB3SP	IO71OQ	RB4	RU248	434.700	433.100	Pembroke	No	GW4VRO
GB3UB	IO81UJ	RB4	RU248	434.700	433.100	Bath	No	G4KVI
GB3VE	IO84SQ	RB4	RU248	434.700	433.100	Great Dunfell	No	G0IYQ
GB3EB	JO01DO	RB5	RU250	434.725	433.125	Brentwood	110.9	G6IFH
GB3GH	IO81WU	RB5	RU250	434.725	433.125	Cheltenham	No	G3LVP
GB3HY	IO90WX	RB5	RU250	434.725	433.125	Haywards Heath	88.5	G3XTH
GB3IM	IO74SG	RB5	RU250	434.725	433.125	Douglas	No	GD3LSF
GB3OV	IO92WD	RB5	RU250	434.725	433.125	Huntingdon	94.8	G8LRS
GB3WJ	IO93QN	RB5	RU250	434.725	433.125	Scunthorpe	No	G3TMD
GB3BD	IO92RA	RB6	RU252	434.750	433.150	Ampthill, Beds	77	G8MGP
GB3BR	IO90WT	RB6	RU252	434.750	433.150	Brighton	No	G4PAP
GB3CR	IO83LC	RB6	RU252	434.750	433.150	Wrexham	No	G8UEK
GB3CW	IO82HL	RB6	RU252	434.750	433.150	Powys	No	GW4NQJ
GB3DI	IO91IN	RB6	RU252	434.750	433.150	Didcot	No	G8CUL
GB3HA	IO93WT	RB6	RU252	434.750	433.150	Hornsea	No	G4YTV
GB3HC	IO82PB	RB6	RU252	434.750	433.150	Hereford	No	G0JWJ
GB3LW	IO91WM	RB6	RU252	434.750	433.150	London	82.5	G7OMK
GB3ME	IO92JJ	RB6	RU252	434.750	433.150	Rugby	No	G0JEW
GB3SK	JO01MH	RB6	RU252	434.750	433.150	Canterbury	No	G6DIK
GB3SY	IO93GN	RB6	RU252	434.750	433.150	Barnsley	71.9	G4LUE
GB3WG	IO81AO	RB6	RU252	434.750	433.150	Swansea	No	GW3VPL
GB3BL	IO92SD	RB7	RU254	434.775	433.175	Bedford	77	G8MGP
GB3DE	JO02NF	RB7	RU254	434.775	433.175	Ipswich	110.9	G1NRL
GB3HZ	IO91QP	RB7	RU254	434.775	433.175	Amersham	No	G0RDI
GB3MF	IO83WG	RB7	RU254	434.775	433.175	Macclesfield	No	G0AMU
GB3MG	IO81EM	RB7	RU254	434.775	433.175	Bridgend	No	GW3RVG
GB3MS	IO82VE	RB7	RU254	434.775	433.175	Worcester	No	G7WIG
GB3NM	IO92KX	RB7	RU254	434.775	433.175	Nottingham	No	G2SP

APPENDIX 2: VOICE REPEATER LISTS

Callsign	Locator	Old chan	New chan	Rx freq (MHz)	Tx freq (MHz)	Site	CTCSS	Keeper
GB3TS	IO94KN	RB7	RU254	434.775	433.175	Middlesbrough	No	G8MBK
GB3WY	IO93BS	RB7	RU254	434.775	433.175	Halifax	No	G8NWK
GB3AN	IO73UJ	RB8	RU256	434.800	433.200	Amlwch	No	GW6DOK
GB3CM	IO71VW	RB8	RU256	434.800	433.200	Carmarthen	No	GW0IVG
GB3EA	IO90HW	RB8	RU256	434.800	433.200	Eastleigh	71.9	G4MYS
GB3EH	IO92FC	RB8	RU256	434.800	433.200	Banbury	No	G4OHB
GB3KV	IO75TW	RB8	RU256	434.800	433.200	Glasgow	No	GM7OLA
GB3LA	IO93ET	RB8	RU256	434.800	433.200	Leeds	No	G8ZXA
GB3PY	JO02AF	RB8	RU256	434.800	433.200	Cambridge	77	G4NBS
GB3TF	IO82SQ	RB8	RU256	434.800	433.200	Telford	No	G3UKV
GB3BE	JO02IF	RB9	RU258	434.825	433.225	Bury St Edmunds	110.9	G8KMM
GB3CL	JO01OT	RB9	RU258	434.825	433.225	Clacton	103.5	G7HJK
GB3CV	IO92GJ	RB9	RU258	434.825	433.225	Coventry	No	G3ZFR
GB3HD	IO93BP	RB9	RU258	434.825	433.225	Huddersfield	82.5	G1FYS
GB3PG	IO75OW	RB9	RU258	434.825	433.225	Greenock	No	GM4PLM
GB3SW	IO91BB	RB9	RU258	434.825	433.225	Salisbury	71.9	G4SXQ
GB3AW	IO91HH	RB10	RU260	434.850	433.250	Newbury	No	G8DOR
GB3BS	IO81RL	RB10	RU260	434.850	433.250	Bristol	No	G4SDR
GB3DD	IO86ML	RB10	RU260	434.850	433.250	Dundee	No	GM4UGF
GB3DY	IO93FB	RB10	RU260	434.850	433.250	Derby	71.9	G3ZYC
GB3LI	IO83LL	RB10	RU260	434.850	433.250	Liverpool	No	G3WIC
GB3LT	IO91SV	RB10	RU260	434.850	433.250	Luton	77	G6OUA
GB3ML	IO85BU	RB10	RU260	434.850	433.250	Airdrie	No	GM3SAN
GB3MW	IO92FH	RB10	RU260	434.850	433.250	Leamington Spa	No	G6FEO
GB3NS	IO91VH	RB10	RU260	434.850	433.250	Reigate	82.5	G0OLX
GB3PB	IO92UO	RB10	RU260	434.850	433.250	Peterborough	77	G1ARV
GB3WO	IO91FU	RB10	RU260	434.850	433.250	Witney	118.8	G4GUN
GB3AH	JO02KP	RB11	RU262	434.875	433.275	Swaffham	No	G8PON
GB3BK	IO91KI	RB11	RU262	434.875	433.275	Reading	118.8	G8DOR
GB3DC	IO94JR	RB11	RU262	434.875	433.275	Sunderland	No	G6LMR
GB3GR	IO92QW	RB11	RU262	434.875	433.275	Grantham	No	G4WFK
GB3GY	IO93XN	RB11	RU262	434.875	433.275	Grimsby	No	G1BRB
GB3HN	IO91VW	RB11	RU262	434.875	433.275	Hitchin	No	G4LOO
GB3HT	IO92HM	RB11	RU262	434.875	433.275	Hinckley	No	G4ALB
GB3LR	JO00AS	RB11	RU262	434.875	433.275	Newhaven	No	G7PUV
GB3RE	JO01HH	RB11	RU262	434.875	433.275	Maidstone	103.5	G4AKQ
GB3RH	IO80MS	RB11	RU262	434.875	433.275	Axminster	No	G6WWY
GB3WP	IO83XL	RB11	RU262	434.875	433.275	Hyde	No	G6YRK
GB3ZI	IO82XT	RB11	RU262	434.875	433.275	Stafford	No	G1UDS
GB3EE	IO93GE	RB12	RU264	434.900	433.300	Chesterfield	71.9	G6SVZ
GB3GB	IO92BN	RB12	RU264	434.900	433.300	Great Barr	67.1	G8NDT
GB3GF	IO91RF	RB12	RU264	434.900	433.300	Guildford	No	G4EML
GB3HM	IO94HB	RB12	RU264	434.900	433.300	Boroughbridge	No	G0RHI
GB3MT	IO83RO	RB12	RU264	434.900	433.300	Bolton	82.5	G8NSS
GB3OX	IO91JS	RB12	RU264	434.900	433.300	Oxford	118.8	G4WXC
GB3PT	IO92XA	RB12	RU264	434.900	433.300	Royston	No	G4NBS
GB3WB	IO81MI	RB12	RU264	434.900	433.300	Weston Super Mare	77	G4SZM
GB3CA	IO84OT	RB13	RU266	434.925	433.325	Carlisle	No	G4KFN
GB3CY	IO93KY	RB13	RU266	434.925	433.325	York	No	G4FUO
GB3DS	IO93KH	RB13	RU266	434.925	433.325	Worksop	No	G3XXN
GB3GU	IN89RL	RB13	RU266	434.925	433.325	St Peter Port, CI	No	GU4EON
GB3HW	JO01CN	RB13	RU266	434.925	433.325	Romford	No	G4GBW
GB3LC	JO03AI	RB13	RU266	434.925	433.325	Louth, Lincs	71.9	M5ZZZ

Callsign	Locator	Old chan	New chan	Rx freq (MHz)	Tx freq (MHz)	Site	CTCSS	Keeper
GB3SM	IO93AC	RB13	RU266	434.925	433.325	Leek	No	G8DZJ
GB3VH	IO91VT	RB13	RU266	434.925	433.325	Welwyn Garden City	No	G4THF
GB3XX	IO92KG	RB13	RU266	434.925	433.325	Daventry	No	G1ZJK
GB3AB	IO84WD	RB14	RU268	434.950	433.350	Aberdeen	No	GM0GIB
GB3CB	IO92BL	RB14	RU268	434.950	433.350	Birmingham	67.1	G8AMD
GB3CE	JO01KV	RB14	RU268	434.950	433.350	Colchester	103.5	G7BKU
GB3ED	IO85JW	RB14	RU268	434.950	433.350	Edinburgh	No	GM4GZW
GB3GL	IO85WL	RB14	RU268	434.950	433.350	Glasgow	No	GM3SAN
GB3HE	JO00HV	RB14	RU268	434.950	433.350	Hastings	103.5	G4FET
GB3HR	IO91TO	RB14	RU268	434.950	433.350	Harrow	No	G1NOC
GB3LF	IO84OA	RB14	RU268	434.950	433.350	Lancaster	No	G3VVT
GB3MR	IO83XH	RB14	RU268	434.950	433.350	Stockport	No	G8LZO
GB3ND	IO70WX	RB14	RU268	434.950	433.350	Bideford	No	G4JKN
GB3SD	IO80SQ	RB14	RU268	434.950	433.350	Weymouth	No	G0EVW
GB3TL	IO92WS	RB14	RU268	434.950	433.350	Spalding	No	G0UOQ
GB3WF	IO93DV	RB14	RU268	434.950	433.350	Otley Chevin	No	G0NIG
GB3YL	JO02UL	RB14	RU268	434.950	433.350	Lowestoft	94.8	G4RKP
GB3FN	IO91OF	RB15	RU270	434.975	433.375	Farnham	82.5	G4EPX
GB3HB	IO70OI	RB15	RU270	434.975	433.375	St Austell	No	G3IGV
GB3LH	IO82OP	RB15	RU270	434.975	433.375	Shrewsbury	No	G3UQH
GB3OM	IO64JQ	RB15	RU270	434.975	433.375	Omagh	No	GI4SXV
GB3PP	IO83PS	RB15	RU270	434.975	433.375	Preston	No	G3SYA
GB3SG	IO81LR	RB15	RU270	434.975	433.375	Pontypool	No	GW8ERA
GB3SU	JO02JA	RB15	RU270	434.975	433.375	Sudbury	No	G8LTY
GB3SZ	IO90BR	RB15	RU270	434.975	433.375	Bournemouth	No	G0API
GB3TH	IO92DP	RB15	RU270	434.975	433.375	Tamworth	No	G4JBX
GB3WI	JO02CP	RB15	RU270	434.975	433.375	Wisbech	No	M0DUQ
GB3WU	IO93EP	RB15	RU270	434.975	433.375	Wakefield	No	G0COA
GB3UO	IO82LX			438.425	430.825	Wrexham	No	G4UDE
GB3PZ	IO83XL			438.500	430.900	Manchester	82.5	G4ZPZ
GB3FJ	JO03CD			438.550	430.950	Lincolnshire	71.9	G8LXI
GB3BH	IO91TP	RM0		1291.000	1297.000	Watford	82.5	G7LXP
GB3MC	IO83RO	RM0		1291.000	1297.000	Bolton	82.5	G8NSS
GB3NO	JO02PP	RM0		1291.000	1297.000	Norwich	94.8	G8VLL
GB3FM	IO91OF	RM2		1291.050	1297.050	Farnham	118.8	G4EPX
GB3CP	IO91VD	RM3		1291.075	1297.075	Crawley	No	G3GRO
GB3PS	IO92XA	RM3		1291.075	1297.075	Royston	77	G4NBS
GB3SE	IO83WA	RM3		1291.075	1297.075	Stoke on Trent	103.5	G8DZJ
GB3CN	IO92NF	RM5		1291.125	1297.125	Northampton	77	G6NYH
GB3BW	IO92SD	RM6		1291.150	1297.150	Bedford	77	G1BWW
GB3MM	IO82XP	RM6		1291.150	1297.150	Wolverhampton	67.1	G4OKE
GB3CO	IO92PM	RM8		1291.200	1297.200	Corby	No	G8MLA
GB3UY	IO93LW	RM13		1291.325	1297.325	York	118.8	G7AUP
GB3WC	IO93EO	RM15		1291.375	1297.375	Wakefield	No	G0COA

APPENDIX 2: VOICE REPEATER LISTS

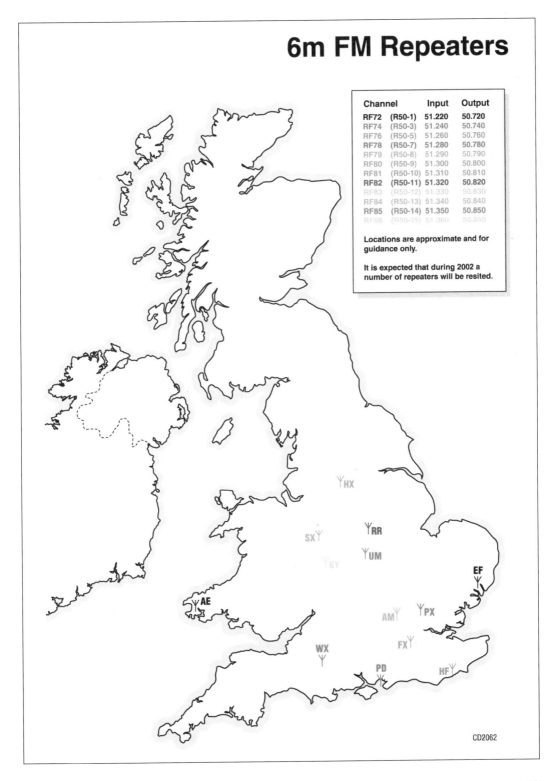

AMATEUR RADIO MOBILE HANDBOOK

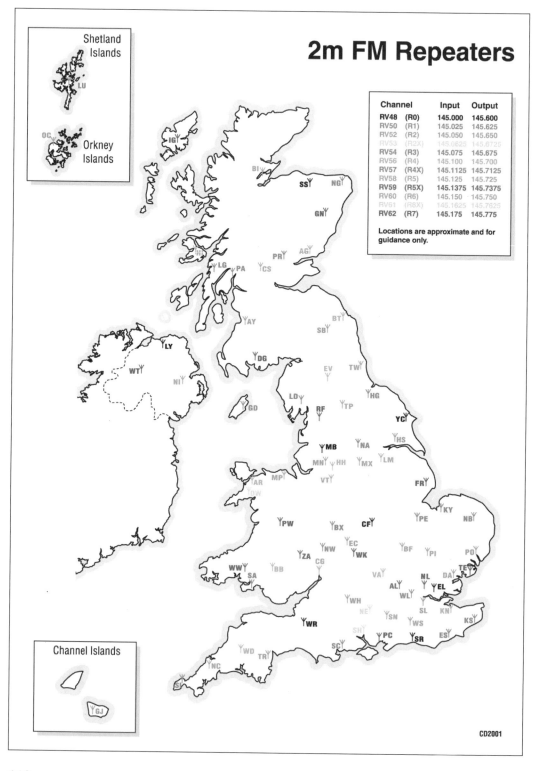

APPENDIX 2: VOICE REPEATER LISTS

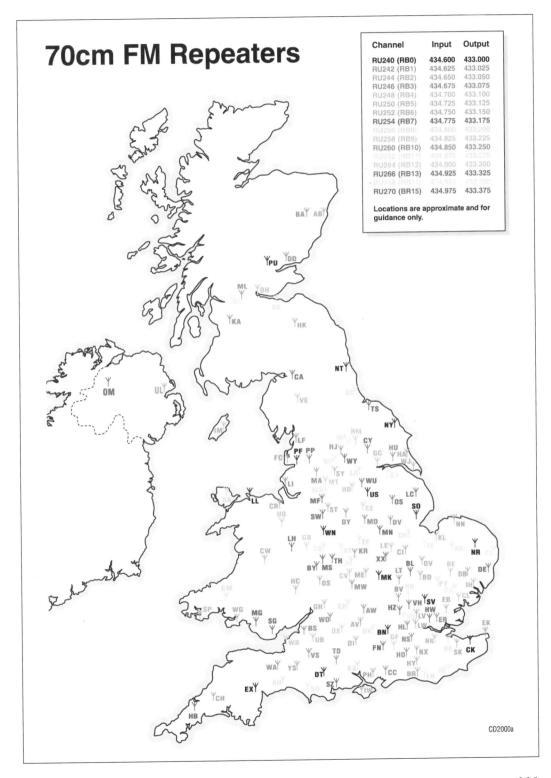

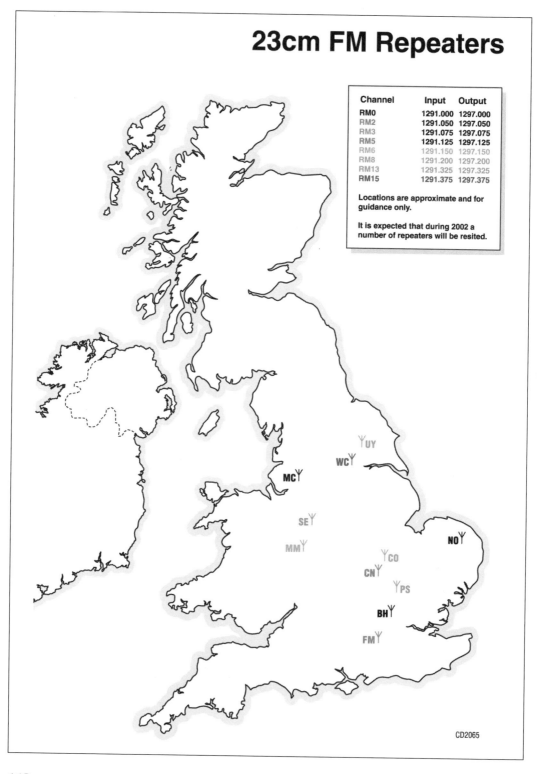

Index

A
Adventure radio 87
Air traffic control regulations 61
Airdux coils 89
Alternator
    suppression capacitors 27
    whine 27
Antennas
    army whip 49
    computer model 32, 52
    current distribution 33
    early mobile design 5
    fixing to vehicle 51
    location 51
    mobile 31*ff*
APRS 15–19
ATU for kites and balloons 64

B
B&W coil stock 36
Backstay antenna 72–74
Balloons 1, 62*ff*
Battery 7, 65, 67, 69
    lead-acid 7
Bicycles 65*ff*
Boats 71*ff*
Bonding, anti-electrolysis 74, 77
Boom microphones 7
BR68 (REV8) September 2000 1
Bull-bar antenna support 58

C
Capacitive coupling 57
Capacitive shunt feeding 45
Capacity hat 32
CB Firestick antenna 38
Centre console, removal 22

Chassis bracket 54, 55
Cigar lighter socket 24
Coaxial cable antenna feeder 59
Coil, loading 35, 36
Computer model, antenna 32, 52
Continuous loading (helical) . . 36, 65
Counterpoise
    marine 74
    pedestrian mobile 89
CTCSS tone 12, 13
CW operating 6, 70

D
Diplexer, antenna 48
DDRR antenna 79, 80, 81

E
Early days 2
Early mobile operation 5
Earth system 74
Engine control unit radiation 27
EMC
    marine 77*ff*
    vehicle 26*ff*
Equipment ergonomics 9

F
Filters, power lead 29
FT-140-43 ferrite cores 29
FT-70G 88
FT-817 91–97
Fuse ratings 24

G
G0CBM 88
    antenna 88, 89, 90
G0TRT 15

G2AHL 3, 4
G2AJV 82
G2ATK/M 2
G2HCG 4
G3AQC 71
G3JKV 3
G3LDO
   bikemobile 65ff
   loading coil 35
G3MPO 32
   base 55
   coil and antenna 33, 34
G3MY 2
G3OJV 40, 41, 42
G3ROO 23
G3SJX 88
G3TSO 32, 47
G3WFM 84
G3WW 4
G3XC 4
G3XDV 85
G3YXM 29, 39, 59, 69
   'Vaerial' antenna 59, 60
G4NXG 9
G5BM 3
G5CP 7, 8
G5CV 3
G5KW 4
G6AG 4
G6XN 37
G8ENN 37
Gamma rod matching 84, 85
Gas, balloon 63
GB2RS 84
GPS 18
G-whip 37

H

Handheld microphones 7
Hatchback antenna mount 58
Helical wound antenna 65
HF operation 9ff
Highway Code 7
History, amateur mobile radio ... 1ff

I

IC-706 .... 65, 87, 90, 92
IC-781 84
IC-M270 transceiver 72
Ignition interference 26
Impulse (noise) limiters 26

Index Labs QRP+ 67
Inductive loading 31
Inductive shunt feed matching... 46
Interference to reception 26ff

K

KB8U bikemobile 69
Kenwood TH-D7E 18
Keyers
   G3LDO lightweight paddle.... 66
   KB8U home-made 69
Kites 1, 61, 62, 63
   Delta kite 61
   ripstop nylon 61
   winder 63

L

Limiters, impulse noise 26
Location, transceiver 21
Logbook, steering wheel 9, 10
Luggage/roof-rack antenna
   support 56

M

Magnetic coupler............... 57
Magnetic mount 51, 79
Manpack 87
Maritime operation 71ff
   ATUs 73, 74, 75
   centre-fed antenna 76
   fibreglass whips 76, 77
   radio installation 71
   RF problems 77, 78
   VHF 78
   VHF masthead antennas 78
Matching to the feeder 44ff
   capacitive shunt feeding 45
   capacitor matching box
     (MFJ-910) 42, 43
   gamma rod 84, 85
   inductive shunt feed 46
   non-resonant antenna 45
   omega match 84, 85
   transformer matching 46
Microphones 7
Mobile antennas 31
'Mobile Column' 2, 3, 7
Mobile rallies 3
MOT test 55
Mounting, through-panel 54

INDEX

N

N7LYY antenna 40
Noise
   alternator whine 27
   ignition interference 26
   impulse noise limiters 26

O

Omega match 84, 85
Operation
   moving vehicle 9
   stationary vehicle 10
Outbacker Perth antenna 67

P

Pedestrian mobile 87*ff*
Portable operation. 87*ff*
Power supply, 12V 24
ProAm antenna 37

Q

Q-TEK five-band HF mobile
       antenna 37

R

Radiocommunications Agency. . . . 1
Radio Club of East Africa 5
Radio installation 21
Repeaters
   abuse 14
   CTCSS tones 12, 13
   frequencies 13
   lists 103–108
   maps 109–112
   operation. 12*ff*
   technology 12
   use 14
Resonant lengths 31
RG174 cable 60
Routing cables in vehicles 25

S

Safety 5*ff*
   batteries, lead-acid 7
   bicycle, 67
   boom microphones 7
   carbon monoxide. 8
   Department of Transport letter . 7

   driving, distraction 6
   handheld microphones 7
   mobile CW operating 6
   UK Highway Code 7
Site, choosing (kites & balloons). . 63
Steering wheel logbook 9, 10
Suppliers 99–101

T

Texas Bugcatcher. 32, 60, 91
Through-panel mounting 54
TM-D700E. 18
Toneburst, repeater 12
Toroidal antenna 82
Transceiver location 21
Transformer matching 46

V

'Vaerial' antenna 43, 44
VE3JC bikemobile 67*ff*
Vehicles
   fixing an antenna to 51*ff*
   handbook 21
   installing radio equipment in . 21*ff*
   routing cables in 25
   warranty 21
VHF/UHF 11
   antennas 47–49
   channel spacing 11
VK6QG 22
VQ4HX/M 6
Voltmeter 10

W

W2WAM 80
W3WAM's 'Viking Mobile' 4
W6AAQ continuous coverage HF
       antenna 39
WB4APR 15
WBB-3 antenna 40, 42
Webster Band Spanner
       antenna 39, 40
W-ECH cable kit 60
Wind (Beaufort scale) speed 63
WMM-3401. 53

Z

ZC1 MKII transceiver 3

MORE BOOKS FROM THE RSGB

HF Antenna Collection

Edited by Erwin David, G4LQI

An invaluable collection of the outstanding articles and short pieces that were published in the Radcom magazine during the period 1968-89. Includes ingenious designs for single element, beam and miniature antennas, as well providing comprehensive information about feeders, tuners, baluns, testing, modelling, and how to erect your antenna safely.

1st Edn, 1992, RSGB, paperback, 184 by 245 mm, 233 pages, ISBN: 1-872309-08-9.

Price: £9.99

The Antenna File - NEW

The Radio Society of Great Britain produces some of the best works on antennas and this is a collection of that work from the last ten years. This book contains 288 pages of articles drawn from the Radcom magazine and includes: · 50 HF antennas, 14 VHF/UHF/SHF antennas, 3 receiving antennas, · 6 articles on masts and supports, · 9 articles on tuning and measuring. · 4 on antenna construction. · 5 on design and theory · And 9 Peter Hart antenna reviews. · Every band from 73kHz to 2.3GHz · Beams, wire antennas, verticals, loops, mobile whips and the G2AJV Toroid. In fact everything you need to know about antennas and how to get the best out of them.

1st Edn, 2001, RSGB, paperback, 297 by 210 mm, 288 pages, ISBN: 1-872309-72-0.

Price: £18.99

HF Antennas for all Locations

By Les Moxon, G6XN

This is a thought-provoking book, which has been a major contribution to the state of the art from an acknowledged expert. It explains the 'why' as well as the 'how' of HF antennas, and takes a critical look at existing designs in the light of the latest developments. This second edition has been completely revised and greatly expanded. There are more novel antenna designs, including beams which cover more bands with fewer problems, no trap losses and better rejection of interference. A new chapter presents a comprehensive review of ways to make antennas smaller, with particular emphasis on small transmitting loops. An essential reference for the experimenter and enthusiast.

2nd Edn, 1993, RSGB, paperback, 187 by 245 mm, 322 pages, ISBN: 1-872309-15-1.

Price: £7.99

www.rsgb.org/shop Tel: 0870 904 7373

RSGB ORDER FORM

ORDERED BY

ORDER NO.   DATE

DELIVER TO

Code	Description	Price	Qty	Total
1-872309-54-2	Backyard Antennas	£18.99		
1-872309-08-9	HF Antenna Collection	£9.99		
1-872309-72-0	The Antenna File NEW	£18.99		
1-872309-15-1	HF Antennas for all Locations	£7.99		
1-872309-11-9	Practical Antennas for Novices	£7.99		
1-872309-36-4	The Antenna Experimenter's Guide	£17.99		
1-872309-74-7	RSGB Yearbook 2002 NEW (available Sept)	£15.99		
1-872309-53-8	Radio Communication Handbook	£29.99		
1-872309-65-8	Low Frequency Experimenter's Handbook NEW	£18.99		
1-872309-40-2	PMR Conversion Handbook	£16.99		
1-872309-30-5	Radio Data Reference Book	£14.99		
1-872309-35-6	Practical Receivers for Beginners	£14.99		
1-872309-21-6	Practical Transmitters for Novices	£16.99		
1-872309-23-2	Test Equipment for the Radio Amateur	£12.99		
1-872309-61-3	Technical Topics Scrapbook 1995-99	£14.99		
1-872309-51-8	Technical Topics Scrapbook 1990-94	£13.99		
1-872309-20-8	Technical Topics Scrapbook 1985-89	£9.99		
1-872309-71-2	The RSGB Technical Compendium NEW	£17.99		
0-705652-1-44	Radio & Electronics Cookbook NEW	£16.99		
1-872309-73-9	Low Power Scrapbook NEW	£12.99		
1-872309-00-3	G-QRP Circuit Handbook	£9.99		
0-900612-89-4	Microwave Handbook **Volume 1**	£11.99		
1-872309-01-1	Microwave Handbook **Volume 2**	£18.99		
1-872309-12-7	Microwave Handbook **Volume 3**	£18.99		
1-872309-48-8	The RSGB Guide to EMC	£19.99		
1-872309-58-5	Guide to VHF/UHF Amateur Radio NEW	£8.99		
1-872309-42-9	The VHF/UHF Handbook	£19.99		
1-872309-63-1	Amateur Radio Operating Manual NEW	£24.99		
N/A	Prefix Guide (fifth edition, 1999)	£8.99		
1-872309-43-7	Your First Amateur Station	£7.99		
1-872309-62-3	The RSGB IOTA Directory	£ 9.99		
1-872309-31-3	Packet Radio Primer	£9.99		
1-872309-38-0	Your First Packet Station	£7.99		
1-872309-49-6	Your Guide to Propagation	£9.99		
1-872309-60-7	Radio Today – Ultimate Scanning Guide	£19.99		
1-872309-27-5	Novice Licence - Student's Notebook	£4.99		
1-872309-28-3	Novice Licence - Manual For Instructors	£9.99		
1-872309-45-3	Radio Amateur's Examination Manual	£14.99		
1-872309-18-6	RAE Revision Notes	£5.00		
1-872309-19-4	Revision Questions for the Novice RAE	£5.99		
1-872309-26-7	Morse Code for Radio Amateurs	£4.99		
1-872309-50-0	Amateur Radio ~ the first 100 years	£49.99		
0-900612-09-6	World at Their Fingertips	£9.99		
Post & Packing		P&P		
UK only - £1.50 for 1 item £2.95 for 2 or more items		Discount		
Rest of World - £2.00 for 1 item 4.00 for 2 & £0.50 for each extra item		Total		

RSGB, Lambda House, Cranborne Road, Potters Bar, Herts EN6 3JE UK
Tel: 0870 904 7373 Fax: 0870 904 7374 E-mail sales@rsgb.org.uk